THE AGE OF ELECTRONICS

LINCOLN LABORATORY DECENNIAL LECTURES

LLOYD V. BERKNER
EDWARD G. BOWEN
HENDRIK B. G. CASIMIR
IVAN A. GETTING
JOHN R. PIERCE
WILLIAM SHOCKLEY
CHARLES H. TOWNES
STANISLAW M. ULAM

McGRAW-HILL BOOK COMPANY, INC.

THE AGE OF ELECTRONICS

edited by · CARL F. J. OVERHAGE
Professor of Engineering and Director, Lincoln Laboratory
Massachusetts Institute of Technology

1962 New York San Francisco Toronto London

THE AGE OF ELECTRONICS

Library of Congress Catalog Card Number 62-17372

47945

To

F. W. LOOMIS

first Director of the Lincoln Laboratory

PREFACE

This book owes its existence to a discussion in the Lincoln Laboratory Steering Committee during the summer of 1961. The Laboratory had been in operation for ten years, and it was felt that some recognition should be given to the anniversary. In the life of an institution ten years is a short span, but for the group of relatively young scientists and engineers who work at Lincoln Laboratory, ten years is an important segment of a professional career. It seemed a good time to look beyond the preoccupations of the moment at the larger world of electronics outside our walls. We concluded that we should invite a small number of the leaders in our field to come to Cambridge to talk to us about those subjects in electronics with which their own careers had been most closely identified. The lectures were given during the winter of 1961–1962 at the Kresge Auditorium of the Massachusetts Institute of Technology in Cambridge. With only minor changes, the chapters of this book are the written record of the Lincoln Laboratory Decennial Lectures.

Lincoln Laboratory is a research center operated by the Massachusetts Institute of Technology with funds provided by the Federal government. Its primary field of work is electronics. In this field, the activities of the Laboratory are divided into two broad categories: general research and specific projects. *General research* at Lincoln extends over a broad range from solid-state physics through communications, radar, information processing, and systems analysis to radar astronomy and space surveillance. This program provides the foundation for the *specific projects* of the Laboratory, which are related to major national objectives in military defense and in the conquest of space. Among the specific projects, the first major enterprise at Lincoln Laboratory was the design of the SAGE system for continental air defense. More recently, the emphasis has been on defense against ballistic missiles, on strategic communications, and on the electronic environment for space vehicles.

The staff of Lincoln Laboratory includes about five hundred scientists and engineers. These men and women were the core of the audience to whom the Lincoln Decennial speakers addressed their lectures. On those evenings, the purpose of the Lincoln group was not to seek new items of technical information in its areas of specialization. Rather, we hoped, the speakers would convey new insights and fresh perspectives by taking a broad approach and by striving for a proper balance between the sober analysis of past accomplishment and the imaginative suggestion of future challenge. We also expected that these lectures would attract many other interested listeners from the schools and laboratories of the Boston area, and we therefore suggested that the technical level of the lectures be set to appeal to a well-educated nonspecialist group.

The selection of eight subjects to represent the vast complexities of the age of electronics was a matter of much thought and discussion. In the end, we put the emphasis on topics that were close to our own pursuits without too much concern for universal approbation of the balance of our choice.

To the eight Lincoln Laboratory Decennial speakers, whose words have been recorded in this book, I should like once again to convey the gratitude of all men and women at Lincoln Laboratory. In the preparation of the final copy for the publisher I have had the assistance of many staff members at Lincoln Laboratory, among whom I am particularly indebted to Dr. Joseph Mindel, whose competent and devoted attention to all details has shielded me from the usual afflictions of editorial responsibility. Drs. Harry C. Gatos and Robert H. Rediker have made substantial contributions to Chapter 7, and Dr. James W. Meyer to Chapter 8. In addition, I am under a deep obligation to Allen S. Richmond for the efficient skill and unfailing good nature with which he has supervised the physical and social arrangements for these lectures.

C. F. J. O.

CONTENTS

THE AGE OF ELECTRONICS

Carl F. J. Overhage has been Director of the M.I.T. Lincoln Laboratory since 1957 and Professor of Engineering since 1961. During World War II he was a member of the M.I.T. Radiation Laboratory. He is a physicist whose career has included both industrial and military research.

CHAPTER 1

INTRODUCTION

Few readers will open this book without bringing to it a personal awareness of the age of electronics. In our lives today, we are surrounded by the manifestations of that age. Antennas are in the field of vision everywhere: television antennas on our roofs, whips on our automobiles, microwave relays on our hills, radar scanners on our ships. Electronics has given us the freedom and the ease with which we talk to each other over great distances. Electronics provides our entertainment: radio and television broadcasts, phonograph records, sound motion pictures. Without electronics, modern air transportation would be impossible. Wherever rockets are launched into space, electronic facilities are available for communications and control. Less conspicuous visually but equally pervasive throughout modern life are the electronic information-processing techniques of which the high-speed digital computer is the outstanding example.

The numbers that describe the growth of electronics are startling even in a world that has become accustomed to the pace of modern technological development. At the end of 1960, the electronics industry in the United States had achieved an annual sales volume of 10 billion dollars. This was four times the volume of 1950. A writer in *Fortune* remarked (in 1960) that "electronics stocks have been a continuing sensation in Wall Street for the last two years." [1] †

Much of the growth during the last decade has been stimulated by new military applications. Year by year, electronics commands a larger fraction of the country's total investment in weapons. But there has also been a continuing expansion in the older branches of electronics

† Superscript numerals are keyed to References at the end of the chapter.

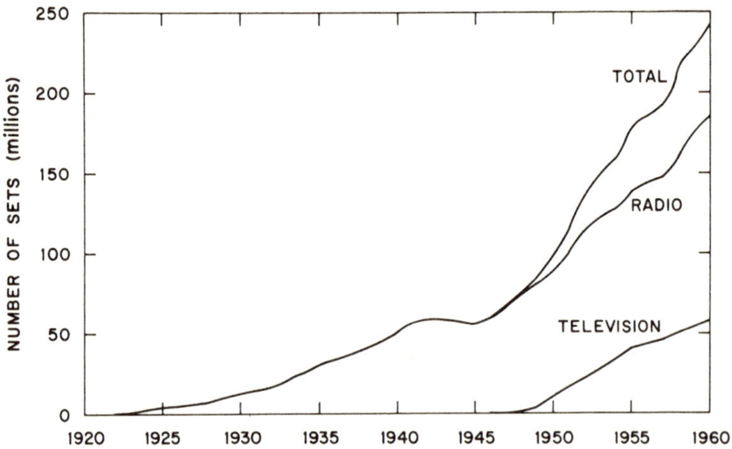

Figure 1. Radio and television sets in use in the United States. (Data from *Electronics Inds.*, vol. 20, no. 1, p. 132, January, 1961.)

manufacture. Figure 1 shows the total numbers of radio and television sets in use in the United States during the last forty years.

The most significant numerical indices of the age of electronics are probably those which pertain to the scientists and engineers whose efforts are the driving force in all this progress. Not every electrical engineer contributes to the advancement of electronics, and certainly not every physicist. Nevertheless, electrical engineering and physics are the major sources of new concepts in electronics. The growth of interest in these two fields is exemplified by the degrees awarded at M.I.T. in the corresponding two departments during the last six decades (Figure 2). An equally impressive and closely related illustration is the record shown in Figure 3 of the membership of the dominant professional society in electronics, the Institute of Radio Engineers.

Many of these scientists and engineers carry on their pursuit of electronics as members of large teams. Electronics will always offer great scope to the outstanding individual, but the bulk of electronics research and development is nowadays done in laboratories that have organized their staffs into integrated teams. The Bell Telephone Laboratories have assembled over four thousand scientists and engineers into a single research and development organization. Under the pressure of wartime urgencies, M.I.T. established its Radiation Laboratory as a temporary organization for the development of microwave radar; at the height of its activities, it had a professional staff of more than one thousand scientists and engineers. The trend away from one-man laboratories toward large aggregations is not unique to electronics; in almost every field of technical endeavor it appears to be the direct counterpart of the increasing complexity of our civilization.

The period which these Lincoln Laboratory Decennial Lectures seek to encompass began on 1 January 1896. That, at any rate, was the date chosen by Robert A. Millikan as the starting point:

I have been in the habit of taking the first of January, 1896, as marking the end of the old period in the development of physics and the beginning of the new. Electronics is the single word which best describes the new period, although the scientific world had not generally accepted the electron and the English interpretation of cathode rays until J. J. Thomson's paper appeared in the *Philosophical Magazine* in 1897. Indeed, Roentgen himself was standing out against "the electron" as late as 1910.[2]

To a whole generation of American physicists, Millikan was more closely identified with the electron than anyone else, even J. J. Thomson. For it was Millikan who, in 1912, had *measured* the charge of the electron, and what better way is there of knowing a thing?

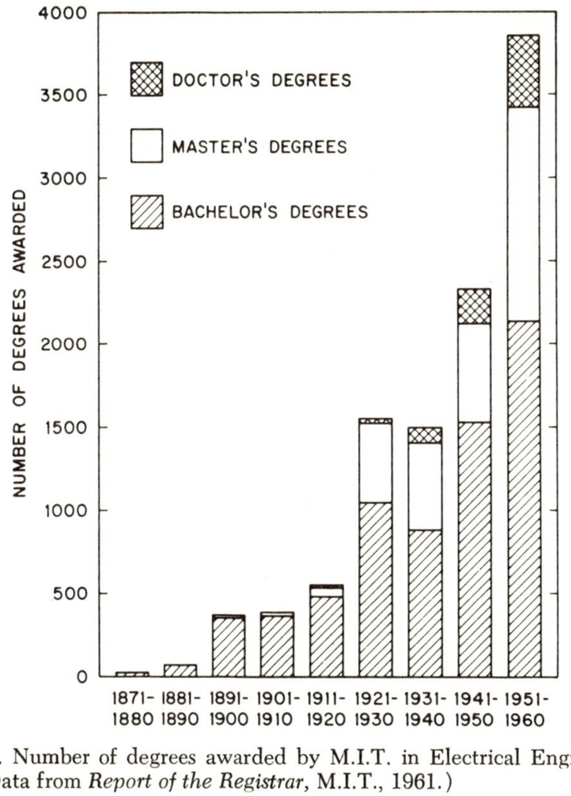

Figure 2. Number of degrees awarded by M.I.T. in Electrical Engineering and Physics. (Data from *Report of the Registrar,* M.I.T., 1961.)

But the assignment of a unique date as the starting point of an age is an uncertain and fanciful business, and the reader may have different thoughts. In Chapter 2, Dr. Casimir has addressed himself to the much more meaningful problem of tracing the great intellectual currents that converged toward the end of the last century to initiate the age of electronics.

The main stream of physics then led on to the great and revolutionary insights of the early twentieth century: Planck's quantum hypothesis in 1900; Einstein's papers of 1905 on the photoelectric effect and on special relativity; Bohr's theory of the hydrogen atom in 1913. Electrons became more and more prominent in the discourse; yet the impact of these advances was slow to reach the public. Physics was a remote subject; science had not yet come to be generally regarded as a dominant factor in shaping the life of our society.

Public awareness of the new age came as a result of a more dramatic event. On a winter night in 1901, Guglielmo Marconi received at New-

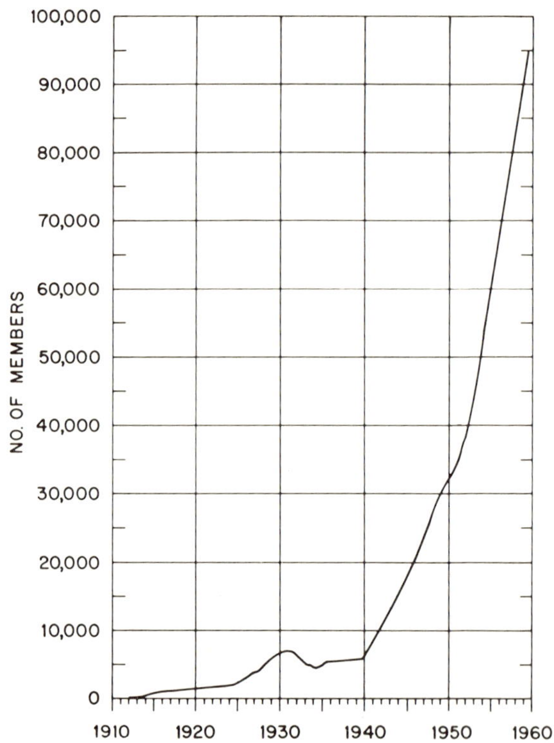

Figure 3. Membership in the Institute of Radio Engineers, 1912 to 1960. (Data from *Proceedings of the IRE.*)

foundland a radio signal transmitted from England. This was a startling climax of successive experiments in a field that has always been linked to one of the most fundamental of human needs: communications. It was clear that the way had been opened toward a tremendous broadening of the scope of communications. Signals would be transmitted to persons at any place on the earth, on the high seas, in the air; to persons whose location was unknown; to many persons simultaneously. All these expectations were fulfilled, and even more exciting horizons came into view as the art progressed. To many people the age of electronics is the age of radio, and it began with Marconi.

But Marconi's experiments were a brilliant exercise in electromagnetics, not in electronics. In any strict interpretation of the matter, the communications field did not enter the age of electronics until 1907 when Lee De Forest invented the triode. In seeking an exact definition for the age of electronics, we are thus caught with the unamiable pedantry of excluding Marconi, if we are to avoid the careless modern tendency of including Samuel Morse and Alexander Graham Bell. In this dilemma it seems best to respect the historical fact that the telegraph and the telephone antedate the age of electronics. The glory of the new age will not suffer, for the tremendous new potentials released for telegraphy and telephony by the introduction of vacuum tubes are too evident to be overlooked or forgotten.

In Chapter 3, Dr. Berkner has lifted his review of communications above these historical particulars by presenting the broadest possible interpretation of his subject. He has taken as his central theme the mathematical generalizations achieved by Shannon [3] and Wiener [4,5] in the 1940s, which established the basis of today's information theory. These new methods have enabled us to obtain a unified view of such diverse communications techniques as telephony, digital data transmission, and television.

The central role of communications in the age of electronics is well exhibited in the history of radar. The earliest observations of radio reflections from ships and aircraft were made by communications engineers, and the crucial technique of pulse ranging was first applied by Breit and Tuve [6] in 1925 to a problem in radio communication: the measurement of the height of the ionosphere. The subsequent exploitation of this technique for military devices was carried on independently in several countries. One of the most inspiring demonstrations of international collaboration in science and technology began in 1940 when Great Britain and the United States decided to consolidate their efforts in radar development. The resulting burst of expansion in research, development, production, and operational use is without parallel, even in the generally spectacular dynamics of the age of electronics. The growth of the child threatened the existence of the father: in 1944, Sir Robert Watson-Watt cautioned the staff of the M.I.T. Radiation Laboratory not to neglect urgent military communications problems in the universal preoccupation with new radar applications.

The stirring historical record of radar in World War II has been given in a number of authentic accounts. Dr. Getting, in Chapter 4, has touched only briefly on that period, so as to give full emphasis to the important and not widely known advances that have been achieved in recent years as radar came to be applied against new threats.

In radio astronomy, we encounter another field, which, like radar, owes its origins to the efforts of communications scientists toward understanding the properties of the transmission medium. Karl Jansky set out to study certain phenomena associated with the operation of long-distance radiotelephone circuits. In 1933, he announced his observations of 20-Mcps electromagnetic radiation of extraterrestrial origin.[7] The full significance of his discovery was not widely appreciated until after World War II. By that time, an enormous reservoir of new techniques and operational experience had become available for exploitation in the design and construction of radio telescopes. In Chapter 6, Dr. Bowen has described an outstanding instrument of this kind, the 210-foot reflector that has recently been constructed under his supervision at Parkes, New South Wales. He has also reviewed the new and fundamental knowledge in astronomy that has come to us from radio observations.

More than any other field of electronics, radio astronomy has been characterized by extensive international cooperation. The centuries-old traditions of the world's astronomers have been extended to the new frequency domain, and radio astronomers of different countries have established a pattern of interaction which is free from the inhibitions of military secrecy and commercial rivalry.

The preelectronic history of computers extends from the obscure origins of the abacus to the desk calculators and many of the analog computers that are in use today. Punched cards, which in the last decade have become so familiar a symbol of modern automation, go back to the last century. Their use in the control of large digital computers was foreseen, as were so many other features of today's large machines, by that remarkable Englishman Charles Babbage, who died in 1871.

But when we speak of the big computers of today, the "giant brains" of the Sunday supplements, we invariably think of the high-speed digital machines that came into existence after World War II. The speed and capacity of these new machines differ by several orders of magnitude from their predecessors. They have opened the way to successful attacks upon entirely new classes of problems. It is this new category of computers which is one of the great triumphs of the age of electronics.

The key to the successful development of these remarkable machines was the use of the pulse techniques that had been worked out in the wartime radar laboratories. The cathode-ray-tube memories in some of the earliest of these machines were another radar-associated component, but in later machines they have generally given way to magnetic storage elements. With either type of memory, the high-speed digital computer is critically dependent on electronic means that have become available only in the last fifteen years. In that short interval, the development of these machines has given rise to a data-processing industry of which the annual sales have now reached a volume of 1 billion dollars.

One of the most significant aspects of this achievement is the impact of the new machines, not only on technology, but on many branches of science. In the problems of pure mathematics, in the advancement of theoretical physics, and also in the biological sciences, new and important progress has become possible by the imaginative exploitation of large computers. The resulting extension of our intellectual domain is the principal topic of Dr. Ulam's discussion, in Chapter 5, of the present status and future prospects of digital computers.

The impact of World War II on the rise of electronics was followed, within the next decade, by revolutionary influences of even greater significance. From Bohr's model of the hydrogen atom, physicists had advanced to a detailed knowledge of the electron structure of atoms and molecules. The methods of quantum mechanics had deepened these insights and had made it possible for theoreticians to come to grips, at last, with the problems of extended structures of atoms and electrons. Solid-state physics had remained a small and recondite domain in the 1930s and 1940s, while the exploration of the nucleus was the central interest in physics. Nevertheless, steady progress had been made, and the combined efforts of theoretical and experimental physicists had gradually led to a detailed understanding of many properties of solid substances.

A particularly interesting class of materials for these studies had been the semiconductors. These are crystals that differ from insulators by the presence of minor imperfections in electronic structure. Thorough study of the electrical behavior of these semiconductors, especially when modified by controlled chemical impurities, pointed the way toward a new electronic amplifying device, the transistor.

The discovery of the first device of this class, the point-contact transistor, was announced by Bardeen and Brattain [8] in 1948. It was followed, in 1949, by Shockley's theoretical discussion [9] of the junction transistor and, in 1951, by its experimental demonstration.[10] In Chapter 7, Dr. Shockley has discussed the advances in our understanding of solid-state physics which are associated with these discoveries.

Several years elapsed before important applications were developed. Thus the period in which transistors have been on the scene as components in electronic circuits is barely a decade. In that period, they have in some fields come to rival vacuum tubes in total importance. The spectacular success of transistors is not wholly attributable to their small size and low power consumption. They have other characteristics that distinguish them functionally from vacuum tubes and permit their use in radically new and different circuits.

There can be no doubt that the transistor is one of the major inventions of our generation. It provides a striking illustration of the changing character of technical invention. Unlike the great inventions of our fathers' generation, the transistor was discovered by the constant interplay of basic laboratory experiments and exhaustive theoretical analysis. The search involved the advanced mathematical resources of modern physics along with the versatile physical and chemical skills of a sizable research organization. This kind of sophisticated team effort has become the prevailing pattern of technological progress in the age of electronics.

The most recent contribution from physics to electronics is the maser. In this device, amplification of microwaves or of optical waves is achieved by induced photon emission accompanied by a transition of the system from a higher to a lower energy state. A variety of means has been employed in different masers to make the upper state continuously available as a source of energy. A substantial field of activities has come into existence for the pursuit of these problems. It is often referred to as quantum electronics. From Dr. Townes's review in Chapter 8, it is clear that this field represents a convergence of quantum mechanical insights, microwave techniques, and new materials.

The first successful demonstration of the principles of maser operation was accomplished by Gordon, Zeiger, and Townes [11] in 1954 with a molecular-beam maser. Two years later, Bloembergen [12] suggested that electron paramagnetic resonance could be used to obtain microwave amplification in certain crystals. Such solid-state masers have been used in constructing low-noise microwave receivers for radio astronomy and for other applications in which extreme sensitivity is required.

The latest and probably the most important development in this field is the optical maser which was proposed by Schawlow and Townes [13] in 1958. The first reports of optical masers in actual operation were given by Maiman [14] in 1960, for a ruby maser, and by Javan, Bennett, and Herriott [15] in 1961, for a helium-neon maser. The discovery of the optical maser has opened the way toward the production of very powerful monochromatic, phase-coherent beams of radiant energy in the visual and infrared regions of the spectrum. The potential applications of such sources are important and varied; they may constitute as rich a technological bonanza as that which was opened by the resonant-cavity magnetron in 1940.

At a time when all the world is preoccupied with man's conquest of space, it would be quixotic to discuss the age of electronics without touching upon the role of electronics in that great adventure. Space vehicles, manned or unmanned, must operate in an electronic environment which includes elaborate facilities for guidance, telemetry, surveillance, and communications. A world-wide network of such facilities has been established by the National Aeronautics and Space Administration for the orbital flights of its Mercury vehicles. Additional equipment is under construction to meet the electronic requirements of lunar exploration. With greater subtlety and over a broader span, electronics will do for the operation of space vehicles what it has long done for aircraft.

There is a more exciting prospect. Man's newly acquired ability to project his equipment hundreds of miles above the earth's surface has given new freedom to his use of radio communications. Since the days of Marconi, intercontinental radio traffic has been limited to frequencies that are reflected by the ionosphere. Modern man is garrulous, and the bandwidth available for long-distance radio circuits in these frequencies is insufficient to meet the demands. Moreover, the quality of the circuits is seriously affected when the upper atmosphere is disturbed by solar phenomena or by nuclear explosions. Radio engineers have long hoped to find a way of gaining both bandwidth and reliability by utilizing the very much higher frequencies that are immune to such interference.

But this could not be done until artificial earth satellites had become a reality. Only then it became possible to put passive reflectors or active repeaters into high orbits so that two rectilinear paths could be joined to form an intercontinental link. Dr. John R. Pierce, in the last chapter of this book, treats of the technical problems associated with this most sensible of all immediate peaceful uses of satellite vehicles.

The topics in this book are large and impressive, but the list falls far short of doing justice to the title. The age of electronics is incredibly rich in stories of human achievement, and the selection of eight subjects to epitomize the age does not reduce the rank of others. Indeed, our daily lives are more directly affected by television, by sound motion pictures, by public address systems, and by hearing aids than by radars or computers. The visitor to Thule in North Greenland is impressed not only by the early warning radars, but also by the "morale circuits" which in the long arctic night enable the soldiers and technicians at that outpost to talk with their families in the United States. The electron microscope, the ultrasonic oscillator, the facsimile transmitter are less prominent in today's technical literature than other subjects, but they are no less significant as manifestations of the age of electronics. Electronic instruments and systems have opened new roads in other fields of science and engineering: radiation counters in medicine, navigation aids in transportation. And is it a negligible achievement to provide time signals throughout the world with an accuracy of a few milliseconds?

More distressing even than the necessary omission of these many technical subjects from this book is the scholarly suppression of the human stories that are part of the greatness of the age of electronics. The lives of Major Armstrong, of Grote Reber, and of Sir Edward Appleton are as moving and as inspiring as any records of high human achievement. And of the purely human images of this age perhaps the most haunting is that of an unknown boy on a remote farm who has built his first amateur transmitter and slowly taps out code to a distant friend whom he has never seen.

REFERENCES

1. C. E. Silberman, *Fortune*, vol. 62, no. 2, p. 126, August, 1960.
2. "The Autobiography of Robert A. Millikan," p. 31, Prentice-Hall, Englewood Cliffs, N.J., 1950.
3. C. E. Shannon, *Bell System Tech. J.*, vol. 27, pp. 379–423, 623–656, 1948.
4. N. Wiener, "Cybernetics," Wiley, New York, 1948.
5. N. Wiener, "Extrapolation, Interpolation, and Smoothing of Stationary Time Series," Wiley, New York, 1949.
6. G. Breit and M. A. Tuve, *Phys. Rev.*, vol. 28, pp. 554–575, 1926.
7. K. G. Jansky, *Proc. IRE*, vol. 21, pp. 1387–1398, 1933.
8. J. Bardeen and W. H. Brattain, *Phys. Rev.*, vol. 74, pp. 230–231, 1948; vol. 75, pp. 1208–1225, 1949.
9. W. Shockley, *Bell System Tech. J.*, vol. 28, pp. 435–489, 1949.
10. W. Shockley, M. Sparks, and G. K. Teal, *Phys. Rev.*, vol. 83, pp. 151–162, 1951.
11. J. P. Gordon, H. J. Zeiger, and C. H. Townes, *Phys. Rev.*, vol. 95, pp. 282–284, 1954.
12. N. Bloembergen, *Phys. Rev.*, vol. 104, pp. 324–327, 1956.
13. A. L. Schawlow and C. H. Townes, *Phys. Rev.*, vol. 112, pp. 1940–1949, 1958.
14. T. H. Maiman, *Nature*, vol. 187, pp. 493–494, 1960.
15. A. Javan, W. R. Bennett, Jr., and D. R. Herriott, *Phys. Rev. Letters*, vol. 6, pp. 106–110, 1961.

Hendrik B. G. Casimir is a theoretical physicist who studied with Ehrenfest, Bohr, and Pauli. He has been Director of the Philips Research Laboratories in Eindhoven, The Netherlands, since 1945, and Professor of Physics at the University of Leyden since 1939.

CHAPTER 2

MAXWELL, HERTZ, AND LORENTZ

1

James Clerk Maxwell was born in Edinburgh on June 13, 1831. His father, member of a distinguished Scottish family and a lawyer by training, did not play a prominent part in public life, but he was intelligent and had a practical turn of mind and a great natural kindness and benevolence.

Maxwell's childhood was passed chiefly at Glenlair, the country home of his father in Kirkcudbrightshire. There he spent his holidays in later life and lived in retirement from 1865 to 1871, working at his "A Treatise on Electricity and Magnetism." It is characteristic of Great Britain that the landed gentry has played a prominent part in its cultural life, and Maxwell was, in many ways, a Scottish country laird. He was courageous, strong, and agile; he dressed for comfort rather than elegance and had a certain Spartan quality about him. He was a good swimmer and an excellent horseman, but he thoroughly disliked hunting: his love for animals was too great. He had a special way with dogs. Once he related in great earnest how he cured a dog of whining at music by simply taking him to the piano and explaining how it worked. This love for animals was in no way a compensation for a dislike of his fellow man but just one facet of a generous and kind-hearted nature. His conversation was witty, stimulating, and slightly eccentric, but he was reticent about his personal affairs. He had a wide knowledge of literature and enjoyed writing verse as a pastime. It is

Figure 1. James Clerk Maxwell.

said that the only language he had any difficulty in mastering was Dutch.

He studied at Edinburgh and at Cambridge, where he became a fellow of Trinity College. From 1857 to 1860, he taught at Aberdeen, and from 1860 to 1865, at Kings College, London. Then, at the age of thirty-four, he left London and from 1865 to 1871 he lived at Glenlair, but in 1871 he returned to Cambridge as the first Cavendish professor, the first of a brilliant sequence of great scientists. He died in 1879.[1]

Heinrich Hertz, born at Hamburg on February 22, 1857, was an entirely different man, a typical member of an intellectual family of Jewish descent—a class which contributed so largely to German culture until Hitler saw fit to discard it, thus giving the final blow to the already dwindling German scientific supremacy.

The father of Hertz was a prominent lawyer and later a senator of the *Hansestadt* Hamburg, and Heinrich received an excellent education. Like Maxwell, he was an outstanding linguist, and as a student he often read Plato and Greek tragedies (in the original, of course) for recreation and general education. He liked to recite Homer and Dante. But other aspects of his education were not neglected: he spent quite a bit of time drawing and modeling and was good at cabinet-making and turning. When the craftsman who had taught him heard about his appointment to a professorial chair, he exclaimed, "What a pity—that boy might have become a real turner!"

Intending to become an engineer, Hertz started with a year of practical apprenticeship with a building firm, but soon after entering the *Technische Hochschule* he decided to become a physicist. From then on, his career was short but brilliant. He obtained his doctor's degree at Berlin in the beginning of 1880, became assistant to Von Helmholtz, was *Privatdozent* at Kiel from 1883 to 1885, professor at Karlsruhe from 1885 to 1889 and at Bonn from 1889 until his death at the age of thirty-seven in 1894. From his numerous letters to his parents (he was always a devoted son), one gains the impression of a high-strung man

Figure 2. Heinrich Hertz.

forcing himself to aim at the highest scientific creativity and quite depressed when the work does not proceed satisfactorily. His enjoyment of nature and literature, though intense, seems less serene, born less out of inner affinity than in the case of Maxwell.[2]

The life of my countryman, Hendrik Antoon Lorentz, was in its outer aspects perhaps even less eventful. He was born at Arnhem on July 18, 1853. His father was a moderately well-to-do gardener, and Lorentz experienced no financial hardships, but his family background did not provide him with any special advantages apart from a tradition of soberness and conscientious work. In 1870, he went to Leyden, where he remained only two years. He returned to Arnhem, where he taught at an evening school and had some difficulty in maintaining order among his pupils. In the meantime, he wrote his doctor's dissertation, obtaining his degree in 1875.

In 1877, he became professor of theoretical physics at Leyden, and there he remained until 1911. He then moved to Haarlem, where a

Figure 3. Hendrik Antoon Lorentz. (The Bettman Archive.)

special position had been created for him by Teyler's Foundation. Nevertheless, he continued to give some lectures at Leyden until his death in 1928.

Lorentz was always a man who liked to work by himself and who did not speak on a subject before it was completely clear in his mind. That explains why he was a most brilliant lecturer and excellent teacher of existing knowledge, knowledge often created by himself, but less successful as a source of inspiration to younger, and less gifted, research workers.

Lorentz had a prodigious memory and an astonishing ease in learning and understanding. He spoke and wrote French, English, and German fluently. He had great charm of manner and considerable diplomatic talent, but he focused his talents conscientiously on his science. Art and literature apparently did not play a prominent role in his life, although his papers are beautifully written and the elegance of his public speeches was proverbial.

Albert Einstein describes him in the following way:

> Just as he mastered physics and mathematical structures, so he mastered also himself, with ease and perfect serenity. His quite extraordinary lack of human weaknesses never had a depressing influence on his fellow men. Everyone felt his superiority; no one felt depressed by it. For, although he had a keen insight into human nature and human relationships, he had a charitable kindness toward it all. His influence was never a dominating, but always a serving, helping one. He was exceedingly conscientious, without, however, making any particular matter appear unduly important. From this, he was protected by his fine sense of humor, which reflected itself in his eyes and in his smile. And it was quite in keeping with this that, with all his devotion to scientific study, he nevertheless was perfectly aware that the human intellect cannot penetrate very deeply into the essential core of things.[3]

2

The discussion of electrodynamic theory will not be strictly historical. A physicist usually makes a poor historian, being always inclined to simplify and modify historical development in such a way that it becomes an image of the way in which ideas have developed in his own mind. He creates a kind of ontogenetic-phylogenetic parallelism by dint of a personal interpretation and selection of historical facts. Every physicist might thus write his own history of contemporary science, but when he discusses past eras, the picture tends to become more and more schematic—less personal perhaps, but even less true. A simplified historical scheme that would be accepted by a majority of physicists is shown in Figure 4.

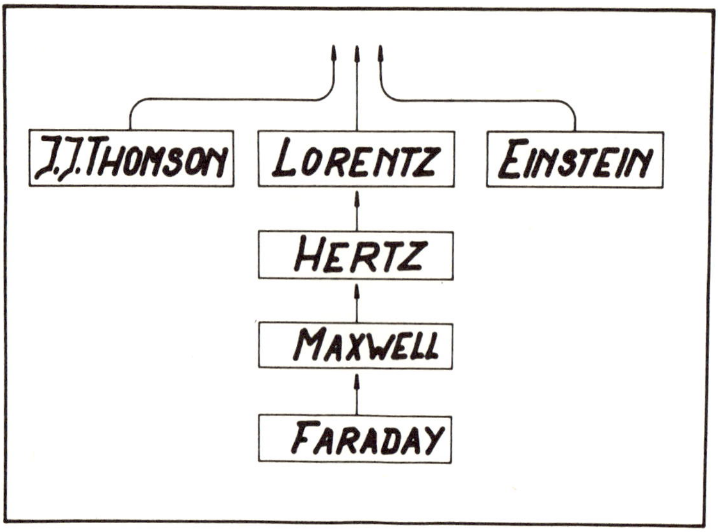

Figure 4. Development of electromagnetic theory: simplified historical outline.

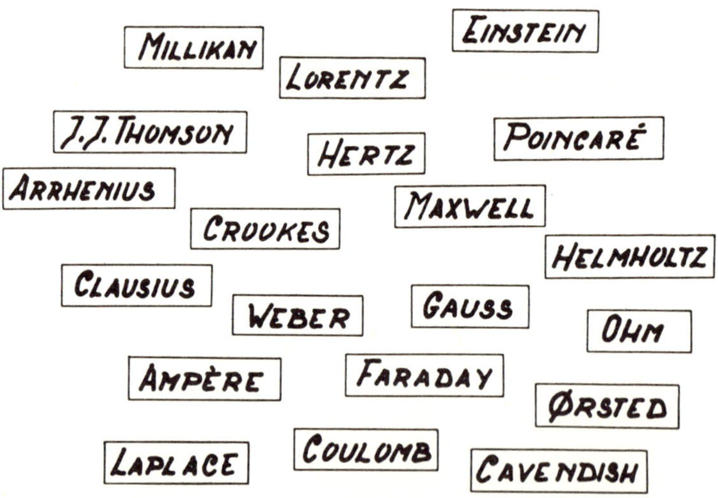

Figure 5. Development of electromagnetic theory: detailed historical outline.

Faraday cleared up most of the basic experimental problems; Maxwell gave Faraday's ideas a mathematical representation, formulated the electromagnetic theory of light, and predicted radio waves; Hertz established the existence of radio waves; and Lorentz incorporated the idea of electrons into the theory and solved the problem of the electrodynamics in moving bodies. Then we need only J. J. Thomson to establish experimentally the existence of the electrons and Einstein to round off the theoretical formalism by introducing the notion of relativity. It is a most satisfactorily simple history—too simple, perhaps, for it omits the important contributions of many scientists, who must be included in a more detailed history (Figure 5). The diagram might be completed by a tangle of arrows.

To quote Lorentz: "Each branch of natural science presents the same picture of related investigations and interwoven ideas."

To try to think oneself back in those days, to look at physics through the eyes of physicists of the past, to point out the ideas floating around, and then to see how the great scientists stand out—not because they are isolated phenomena but just because of their relations with others, not as isolated peaks rising from a plain, but as the highest points of a massif—all this must be left to the historian. We shall consider the extent to which the basic thoughts of contemporary electrodynamic theory are present in, or deviate from, the works of Maxwell, Hertz, and Lorentz.

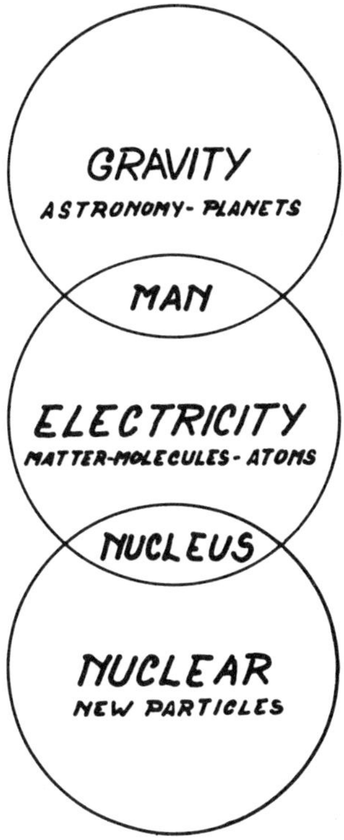

Figure 6. Fields of force in relation to matter and to man.

The place of electricity in natural philosophy may be explained in terms of its relationship to three types of fields of force: gravitational, electromagnetic, and nuclear (Figure 6). The motions of planets and their satellites are ruled by gravity. Electromagnetic forces are not only those that are immediately recognizable as such. In a more sophisticated form, they determine the cohesion of matter, the binding of atoms in molecules, and the structure of the atom, the binding of electrons to the nucleus. But insidc the nucleus we meet with an entirely different field of force between protons and neutrons.

We may say that we are living in the region of intersection of gravity and electricity: it is gravity that keeps my feet on the floor, but electric force that gives the floor its strength. Similarly, in the nucleus, electric force is important as well as nuclear force. In uranium fission, it is electric repulsion that provides the energy. In fusion, on the other hand, the energy is provided by nuclear attraction that has to overcome electric repulsion.

There is in modern theory a pronounced dualism between charges and currents, on the one hand, and electric and magnetic fields, on the other hand. Charge is always associated with particles: negative charge with electrons (-1.6×10^{-19} amp-sec), positive charge with nuclei (1.6×10^{-19} Z amp-sec, where the integer Z is the atomic number). This excludes, of course, the many ephemeral particles of modern physics: positive electrons, negative protons, mesons, etc.

The charge in an element of space may be due to two causes: a real surplus or deficit of electrons, or a displacement of electrons with respect to the positive nuclei, i.e., polarization. It is easy to see that there will be, on the average, a net charge in a region where the polarization changes. Mathematically, this is expressed by the equation

$$\rho_{\text{total}} = \rho_{\text{surplus}} + \rho_{\text{polarization}} = \rho_{\text{surplus}} - \nabla \cdot \mathbf{P}$$

where ρ is the charge density and **P** is the polarization vector.

A current consists of moving charges. But one may distinguish three types of currents: "true" currents, where charges move over considerable distances; displacement currents due to a polarization changing with time; and circulating currents in atoms or molecules associated with magnetism. A number of such currents in atoms or molecules is equivalent to one larger circulating current (Figure 7). We can therefore say that the total current is the sum of convection current + displacement current + magnetic current. In mathematical formulation

$$\mathbf{I}_{\text{total}} = \mathbf{I}_{\text{con}} + \mathbf{I}_{\text{dis}} + \mathbf{I}_{\text{mag}} = \rho \mathbf{v} + \frac{d\mathbf{P}}{dt} + \nabla \times \mathbf{M}$$

where **I** is the current density, **v** is the velocity of the moving charge, and **M** is the magnetization vector.

Incidentally, this decomposition of total charge and total current into different components is by no means always unambiguous. An electromagnetic field manifests itself primarily by its influence on charges

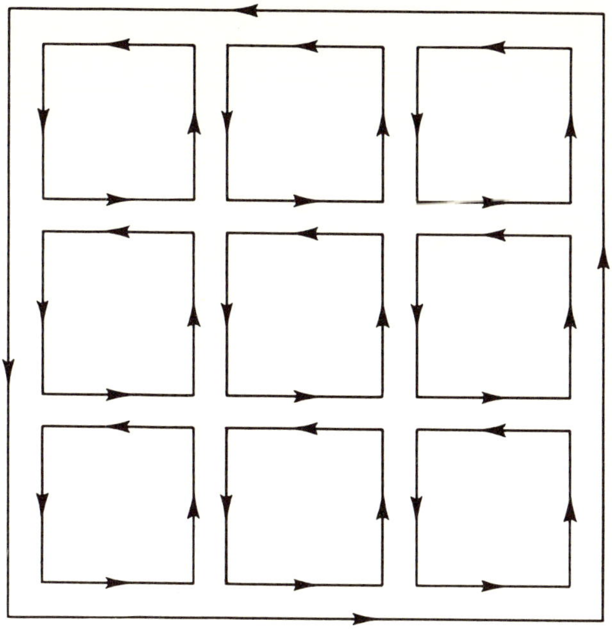

Figure 7. Larger circulating current arising from circulating currents in atoms or molecules.

and currents. An electric field **E** manifests itself by a force $e\mathbf{E}$ on a charge e; a magnetic field **B**, by a force "$\left(\frac{e}{c}\right)$" $\mathbf{v} \times \mathbf{B}$ (the Lorentz force) on a charge moving with velocity **v**. The Lorentz force is at right angles to both velocity and magnetic field; the quotation marks should remind us that the value of the multiplication constant depends on the choice of units.

We talk about electric and magnetic fields and make calculations, as if they were physical realities; we are convinced they are but do not know exactly what we mean when we say so. In any case, while we can determine the existence of a field only by means of its influence on charges and currents (or perhaps, in principle, by some slight gravitational effects), we think there is something there, even in empty space, but this "something" may be no more than a mathematical abstraction.

We come then to the field equations of Maxwell. In words they may be stated as follows: charges and variable magnetic fields produce electric fields; currents and variable electric fields produce magnetic fields (variable meaning variable in time).

Electric lines of force emerge from charges as shown in Figure 8.

The total electric field emerging from a closed surface is proportional to the total charge contained inside that surface. Mathematically, $\oiint E_n\, d\sigma = \text{“}4\pi\text{”} Q$, Q being the total charge, E_n the normal component of the electric field, and $d\sigma$ the element of surface area. The occurrence of the factor 4π depends on the choice of units.

On the other hand, no net magnetic field emerges from a closed surface: there are no magnetic poles (Figure 9), or $\oiint B_n\, d\sigma = 0$.

The relation between electric field and a varying magnetic field is shown in Figure 10. The electric field can circulate around the varying magnetic field. Mathematically,

$$\oint \mathbf{E} \cdot ds = -\frac{1}{\text{“}c\text{”}} \iint \frac{dB_n}{dt}\, d\sigma$$

where ds is the vector element of displacement, $\frac{dB_n}{dt}$ is the rate of change of the magnetic field normal to a surface, and $d\sigma$ is the element of surface area. This equation is the mathematical formulation of Faraday's induction laws.

Finally, the magnetic field circulates around the resultant of the total current and the change of the electric field (Figure 11):

$$\oint \mathbf{B} \cdot ds = \frac{1}{\text{“}c\text{”}} \iint \frac{dE_n}{dt}\, d\sigma + \text{“}\left(\frac{4\pi}{c}\right)\text{”} \iint I_n\, d\sigma$$

where I_n is the current density normal to a surface.

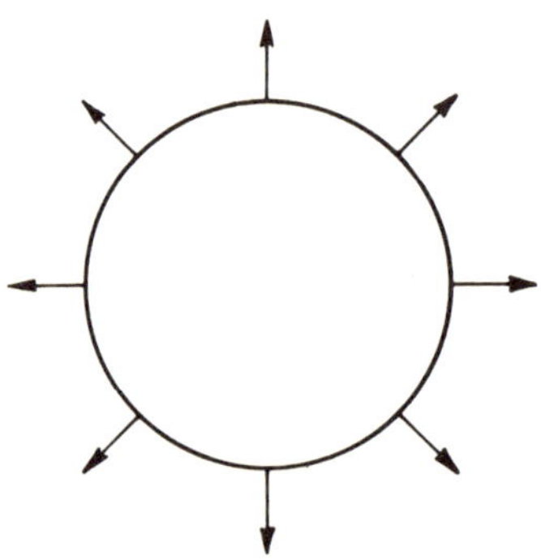

Figure 8. Electric field produced by charge Q contained in a closed surface.

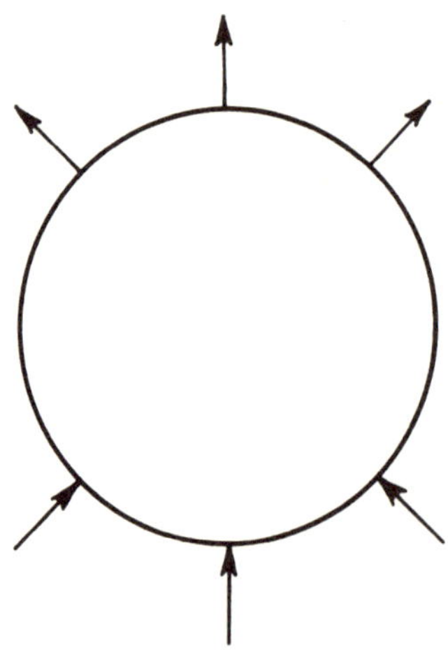

Figure 9. Distribution of magnetic field around a closed surface.

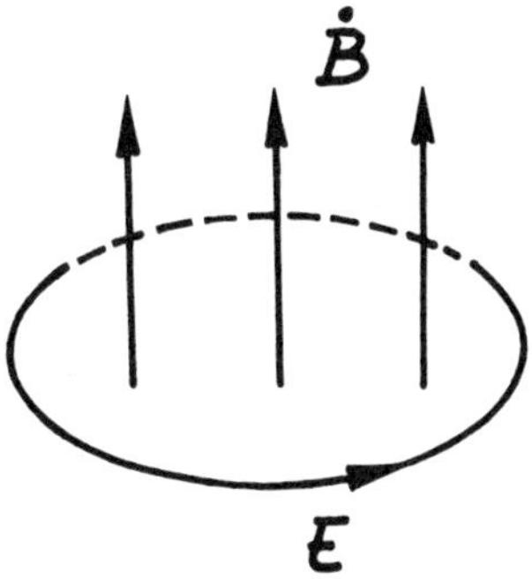

Figure 10. Relation between an electric field and a magnetic field varying with time.

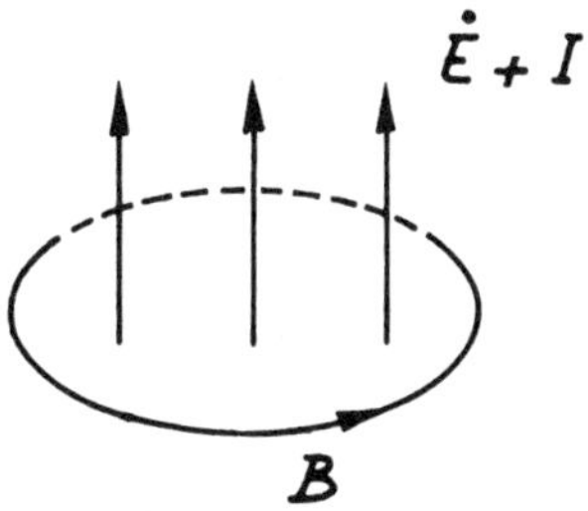

Figure 11. Magnetic field due to a current and an electric field varying with time.

These equations contain a complete description of electromagnetism, but have to be supplemented by equations giving the dependence of currents, polarizations, and magnetic moments on external fields. This dependence may be studied experimentally as well as theoretically by investigating the motion of electrons.

3

How much of this theory do we find in Maxwell? The equations are substantially there, and the greatest advance made by Maxwell was that he recognized a variable electric field as a source of magnetic fields. The existence of electromagnetic waves is then a simple mathematical consequence, but the identification of these waves with light was a bold hypothesis, amply confirmed by subsequent work.

His interpretation, however, was very different. First of all, it is rather surprising that Maxwell, who made outstanding contributions to the kinetic theory of gases, rejected the idea of "a molecule of electricity" suggested by Faraday's laws of electrolysis. In his treatise, he gave a lucid exposition of this idea, but then continued:

This theory of molecular charges may serve as a method by which we may remember a good many facts about electrolysis. It is extremely improbable, however, that when we come to understand the true nature of electrolysis we shall retain in any form the theory of molecular charges, for then we shall have obtained a secure basis on which to form a true theory of electric currents, and so become independent of these provisional theories.[4]

Next, he conceived of electric fields as stresses in a hypothetical medium, the ether. Let me quote two passages. The first one introduces the chapter on the electromagnetic theory of light.

In several parts of this treatise an attempt has been made to explain electromagnetic phenomena by means of mechanical action transmitted from one body to another by means of a medium occupying the space between them. The undulatory theory of light also assumes the existence of a medium. We have now to shew that the properties of the electromagnetic medium are identical with those of the luminiferous medium.

To fill all space with a new medium whenever any new phenomenon is to be explained is by no means philosophical, but if the study of two different branches of science has independently suggested the idea of a medium, and if the properties which must be attributed to the medium in order to account for electromagnetic phenomena are of the same kind as those which we attribute to the luminiferous medium in order to account for the phenomena of light, the evidence for the physical existence of the medium will be considerably strengthened.

But the properties of bodies are capable of quantitative measurement. We therefore obtain the numerical value of some property of the medium, such as the velocity with which a disturbance is propagated through it, which can be calculated from electromagnetic experiments, and also observed directly in the case of light. If it should be found that the velocity of propagation of electromagnetic disturbances is the same as the velocity of light, and this not only in air, but in other transparent media, we shall have strong reasons for believing that light is an electromagnetic phenomenon, and the combination of the optical with the electrical evidence will produce a conviction of the reality of the medium similar to that which we obtain, in the case of other kinds of matter, from the combined evidence of the senses.[5]

The following quotation is the closing paragraph of the treatise.

But in all of these theories the question naturally occurs:—If something is transmitted from one particle to another at a distance, what is its condition after it has left the one particle and before it has reached the other? If this something is the potential energy of the two particles, as in Neumann's theory, how are we to conceive this energy as existing in a point of space, coinciding neither with the one particle nor with the other? In fact, whenever energy is transmitted from one body to another in time, there must be a medium or substance in which the energy exists after it leaves one body and before it reaches the other, for energy, as Torricelli remarked, "is a quintessence of so subtile a nature that it cannot be contained in any vessel except the inmost substance of material things." Hence all these theories lead to the conception of a medium in which the propagation takes place, and if we admit this medium as an hypothesis, I think it ought to occupy a prominent place in our investigations, and that we ought to endeavour to construct a mental representation of all the details of its action, and this has been my constant aim in this treatise.[6]

Even more at variance with modern ideas is his idea of dielectric displacement *in vacuo*. Maxwell imagines that in the ether, too, charges are present and are displaced, just as in a dielectric according to modern theory, but continuously rather than discretely. To Maxwell, then, change of field in empty space *is* really a current.

The word electricity is used by Maxwell in a somewhat ambiguous way: it may refer to surplus charge but also to the hypothetical charges in the ether.

Both Hertz and Poincaré have complained that this makes Maxwell rather hard to understand, not because of the mathematics (that would be rather surprising in the case of Poincaré), but because of the vagueness of the physical ideas. Yet it should at least be admitted that these ideas have had considerable heuristic power.

Hertz reformulated Maxwell's equations, placing far less stress on the idea of an ether. It may be introduced a posteriori to lend more color to the theory, but the equations come first.

"To the question, 'What is Maxwell's theory?' I know of no shorter or more definite answer than the following: 'Maxwell's theory is Maxwell's systems of equations,' "[7] he said in the introduction to the volume in which his papers on electromagnetism were collected.

Hertz made his greatest contribution, not by his theoretical work, but by the series of experiments which established beyond doubt the existence of electromagnetic waves. These will not be described in detail. There is an excellent excuse for referring the reader to the original papers: Hertz himself preferred to study original sources—Lagrange, for instance, and Laplace's "Mécanique Céleste"—rather than textbooks.

Most of the work was done at Karlsruhe between the summer of 1886 and the end of 1888. The results were not at once entirely clear-cut. There were many spurious effects: the mutual influence of spark gaps by ultraviolet radiation; reflections from the walls, especially in the case of longer waves; and there seemed at first to exist a rather baffling difference between the speed of propagation along wires and that in empty space.

The most spectacular paper is that of December 1888.

The emitter consists of a brass rod 26 cm long and 3 cm in diameter, interrupted in the middle by a spark gap, the poles of which are polished spheres of 2 cm radius. During the experiments, the width of the gap is set at 3 mm. The peak voltage produced by an induction coil is connected by insulated wires to the spark gap.

The whole arrangement bears some resemblance to a motor-car ignition system which is now well known to produce radio waves if no special precautions are taken. For detection of waves, Hertz uses another spark gap consisting of a brass sphere a few millimeters in diameter and a sharp point of copper at a distance of a few hundredths of a millimeter, connected to an antenna. The antenna is either a circular loop 7.5 cm in diameter, or it consists of two rods, 50 cm long and 5 mm in diameter, arranged on a straight line with an interruption of 15 cm where a connection to the spark gap is made. Hertz observes the secondary sparks, the intensity of which gives an indication of the strength of the electromagnetic field. Soon afterward, his contemporaries introduced other devices: thermocouples, bolometers, and frogs' legs.

In many of the experiments, emitter and receiver are placed in the focal lines of two large metal mirrors in the shape of parabolic cylinders. This gives considerable "antenna gain," to use a modern expression, and enables him to observe secondary sparks at a distance of 16 meters.

With this equipment, he first of all establishes the existence of nodes and antinodes in front of a reflecting surface and determines the wavelength at 66 cm. He demonstrates rectilinear propagation, with a certain amount of broadening due to diffraction interception by metals, and penetration through insulators. He demonstrates reflection by several experiments: for instance, by throwing the beam, by means of a plane metal mirror set at a 45° angle, through a wooden door into an adjoining room. He studies polarization by turning his mirrors around a horizontal axis and by means of a number of parallel copper wires wound on a wooden frame. Also, he shows refraction by means of a large prism made of pitch.

Many of these experiments were revived after the advent of radar and the modern art of centimetric and millimetric waves.

By his experiments, Hertz proved beyond doubt that a changing electric field creates a magnetic field, and he established the existence of electromagnetic waves propagated with the velocity of light. At the same time, his more formal approach, influenced by the German school of thought, deprived the ether, to a certain extent, of the almost tangible qualities bestowed on it by Maxwell.

The formulation of electromagnetic theory by Lorentz is essentially contemporary, with the difference that Lorentz considers electric and magnetic fields to be manifestations of an ether. This ether he imagines to pervade all matter, even the interior of electrons and atoms, and to remain at rest even inside a moving body. Lorentz was loath to abandon the notion of such a medium altogether, although in reality it had become little more than a kind of mental scaffolding for the formalism of his equations. Whereas in Maxwell's treatise, polarization of a dielectric and polarization of empty space are very much the same kind of thing, this is no longer true with Lorentz. In his theories, Maxwell's displacement current in empty space is something entirely different from a real current. To Albert Einstein, the greatest contributions of Lorentz to the progress of physics were that he did away with the notion of two fundamentally different quantities, electric field and electric displacement, that were required by the older theories to describe an electric field *in vacuo* (and similarly with magnetic field and magnetic induction); and that he pointed out the essential difference between empty space and material media.

Of course, a choice of units may yet render it advantageous to introduce two different symbols for what is essentially one and the same quantity, whereas a generalization of the formalism may make it necessary to distinguish between covariant and contravariant quantities. This does not change the physical situation, but perhaps a word of warning is not out of place. The great, and justifiable, preference of the electrical engineer for the mks system of units may encourage a return to pre-Lorentzian ideas, and thus a modern trend may lead to a retrograde attitude.

Lorentz was able to solve the problem of the electrodynamics of moving bodies. Since the electromagnetic action of a body can be described entirely in terms of charges and currents, it is clear that there should exist an unambiguous answer to what happens when a body is moving. Developing the formalism in the most logical way, Lorentz (and, in a more elegant mathematical form, Poincaré) found a remarkable property of invariance. The equations of electrodynamics in a system that is moving with respect to the ether at rest are identical with the equations in a system that is not moving. The notion of a stationary ether had led to a theory in which motion with respect to this ether is unobservable. The final breakthrough came when Einstein introduced the idea of relativity and abandoned the idea of the ether altogether.

To the modern mind, the tenacity with which the older physicists clung to the notion of a material medium to carry the electromagnetic field seems curious and almost beyond understanding. We have convinced ourselves that matter is composed of molecules or atoms, which themselves are composed of nuclei and electrons, and that the forces between these particles are electrical in nature. In this way, we have reduced the properties of every material medium to electromagnetism; to want to describe electromagnetism in terms of a medium seems like going around in circles. The reason for this difference in outlook is, of course, that we have come to regard atoms and electrons as very real physical entities, not as hypothetical fictions invented for the sole purpose of explaining microscopic properties.

4

Electrodynamics in Lorentz's formulation provides the basis for every branch of electrical and electronic engineering. The importance of these developments is underlined by the fact that it is not necessary to discuss them at all. The electrical theory of matter went through a revolutionary development. By applying the theory of electricity to atomic electrical particles, Lorentz was able to explain many electromagnetic and optical properties of matter, and thus he became one of the pioneers in establishing perhaps the most essential characteristic of modern physics, the quantitative treatment of similar atomic particles possessing accurately known properties. But, although the success of this method must be admitted, there was also a limitation. While, in the beginning, there was doubt as to the reality of the atoms but, at the same time, a most complete confidence in classical mechanics and the deterministic description of space and time that goes with it, the result proved to be different. The arguments for the existence of atoms and electrons became more and more compelling, but ever more threatening at the same time: what Kelvin called the dark clouds gathering on the horizon. It became more and more obvious that the classical methods and notions were not sufficient. And thus a change in the aspect of physics took place. Instead of the exact treatment of a model, to which none too high a degree of reality could be assigned, came the tentative and incomplete description of particles and systems, the reality of which could not be called in question. The quantum theory of Max Planck, the even more advanced ideas of Einstein, the theory of Bohr based on both, and finally the quantum mechanics of Schrödinger, Heisenberg, and Dirac are steps leading toward a more or less complete formalism, which may take the place of classical mechanics and electrodynamics. Lorentz did not contribute to this development to the same extent to which he influenced the first stages. But, within the framework of classical reasoning, he advanced the theory as far as possible and thus created a fresh starting point from which all further theories depart. Is there a contrast between Lorentz, the classical scholar, and the moderns, such as Bohr and his contemporaries? Perhaps to a certain extent this may be admitted. To Lorentz, the reality of the atoms and the electrons turned up as a beautiful result; to Bohr, this reality had become a nearly self-evident fact. But, if this starting point be accepted, the atom, always returning to the same condition, would show a sign of stability unknown to continuous classical mechanics. Then one need no longer regard the failure of classical mechanics as a threatening cloud, but as a self-evident fact which one should take into consideration from the very beginning.

Incidentally, it is interesting to note that Maxwell was deeply impressed by the fact that atoms or molecules on very distant stars emit exactly the same spectra. He can give no other explanation than the assumption that the molecules have been created all alike and are invariable in their properties, and he contrasts this with the development of species of animals. It is this same concern with the stability and permanence of atomic properties that led Niels Bohr to the quantum theory of the atom.

The road to a new mechanics has been a long and difficult one. There may be a difference of opinion as to whether one can apply the term definite to the result reached as yet, especially as far as its interpretation is concerned. It is certain that the result could be reached only by continually using as close an analogy as possible with the classical methods, which remain valid as limiting cases.

One of the problems with which Lorentz too struggled has not been solved yet. In Lorentz's field theory, the structure of the electron remained an enigma. Was it a point? No, for then the field energy would be infinite. Was it a charged sphere? If so, how could the particle stay together while the theory of relativity, the validity of which one surely would want to keep intact, creates additional difficulties? The same difficulties remain principally unsolved in the modern quantum mechanical formulation of the field theory. Their partial solution has led to some wonderful experimental results, but also to a difficult, impenetrable forest of mathematical symbolism.

Beginning as a study of eccentric curiosities, electricity has become a central discipline of physics vying with mechanics for first place. Mysterious action at a distance was replaced by the notion of a concrete medium, but the electromagnetic field, bursting into full bloom, became a mathematical abstraction. Nevertheless, the human mind is capable of handling and applying such abstractions as well as more tangible things: mathematics, no longer a mere tool for solving problems but a fundamental vehicle for our thinking about nature, has almost become one of our senses. The fruitfulness of applying such abstract notions to concrete realizations is clearly demonstrated by the institution whose tenth anniversary we are celebrating this winter.

REFERENCES

1. For further biographical details, refer to L. Campbell and W. Garnett, "The Life of James Clerk Maxwell," Macmillan, London, 1882.
2. Johanna Hertz, "Heinrich Hertz, Erinnerungen, Briefe, Tagebucher," Akademie-Verlagsgesellschaft, Leipzig, 1927.
3. G. L. de Haas-Lorentz, "H. A. Lorentz: Impressions of His Life and Work," pp. 8, 9, North-Holland Publishing Company, Amsterdam, 1957.
4. James Clerk Maxwell, "A Treatise on Electricity and Magnetism," 3d ed., vol. 1, p. 381, Dover, New York, 1954.
5. Ref. 4, vol. 2, p. 431.
6. Ref. 4, vol. 2, p. 493.
7. Heinrich Hertz (D. E. Jones, trans.), "Electric Waves," p. 21, Macmillan, London, 1900.

Lloyd V. Berkner, President, Graduate Research Center of the Southwest, Dallas, Texas, has worked in the field of electronics research and engineering since the early days of radio broadcasting. He is Chairman of the Space Science Board of the National Academy of Sciences and President of the Institute of Radio Engineers.

CHAPTER 3

COMMUNICATIONS

1

An announcement in a *New York Times* editorial symbolizes succinctly the immense influence of today's communications on man's social organization.

"The *Times* announces today a six-day weekday morning edition to be printed in Los Angeles. . . . This development . . . is only made possible by the extraordinary evolution in the science of electronic communications. . . ."

On the basis of the opportunities implicit in today's communications, one can clearly forecast the eventual replacement of our city newspapers, traditional for three centuries, by a few comprehensive dailies of national or even international scope, supplemented perhaps by appropriate local newssheets. This revolutionary social concept derives entirely from the massive capacity for fast, cheap, and precise communications that our technology can now provide. Yet this capability for communication has hardly been tapped.

Webster's International Dictionary defines communications as intercourse by words, letters, or messages; interchange of thoughts or opinions. Consequently, communications in the most fundamental sense involves implicitly a succession of measures.

1. Generation of a thought pattern or image in the human mind
2. Expression or description of that image, with a certain measure of precision, by means of mutually recognizable aural or visual symbols
3. Encoding of these symbols in a form suitable for propagation
4. Propagation of the encoded symbols to the desired destination
5. Decoding and reproduction of the initial symbols

6. Re-creation of the image, with a definable degradation in quality, in the mind of the recipient

Hence communications in its fullest sense involves the transfer of an image from the mind of the originator to the mind of the recipient by a combination of intervening procedures.

There are today many forms of communication not directly involving the human mind in real time. Indeed, most communications today are primarily one-way, and often directly between machines, where human decision enters only in setting up the program or orders for the machine, or in evaluating or monitoring the results.

Data transmission between machines, such as typesetting of the *Times* in Los Angeles from New York; accounting, inventory control, pricing in large industrial complexes, involving collection and processing of data from widespread geographical areas; specifications transmitted directly to the machine, which follows the directions in turning out a special product, even packaging it and preparing it for shipment under distant control—these are all examples of ever-growing unidirectional communication practices.

Navigation through transmission of specialized information directly for control of vehicles, and for monitoring their response, becomes imperative as vehicular speeds increase greatly. Radar, involving a radio echo reflected from the desired object, or a signal triggered from a transponder which responds to enhance the radar signal, is employed in innumerable ways. There is a tendency to forget the wide variety of such powerful and extensive communication applications in our day-to-day personal preoccupation with the telegraph, the telephone, broadcast radio and television, or even the newer closed-circuit conference methods of meetings.

Yet, in the long view, communications must ultimately be responsive to the images of the human mind. Consequently, the efficiency of communications has been greatest when the process of image transfer has conformed as closely as possible to the fundamental processes.

In our day-to-day life, the conveyance of ideas through speech, motions, written words, or in a variety of other ways, is so automatic that we tend to give the process no conscious consideration. But when one reflects on the matter, it is astonishing that through a wholly abstract, and in itself quite meaningless, operation, a complex idea can be transferred with considerable precision from the mind of one person to the mind of another.

The transfer of information or ideas appears to involve two separate processes: the formulation and conveyance of the information, and the process of learning or understanding what that information means. Conveyance may be accomplished by signs or symbols, or by some combination, but understanding requires the application or the extension of previous experience that is common to the persons involved.

In examining the procedure of conveyance, an important distinction must be made between "signs" and "symbols." This difference has been described by the psychologists Yerkes and Nissen.

> The ringing of a bell can, by repetition, act as a sign indicating the approach of mealtime or an electric shock, but it does not represent the meal or shock. If not followed by them, after a time soon it loses that meaning, and ceases to be a sign. But the word "bell" represents that particular object, and continues to do so, even though no bell may be present to the senses for years. The sign is not a substitute, the symbol may be.[1]

The ability to learn is apparently present to some degree in all forms of animal life.

> From single-celled protozoa to animals consisting of millions of millions of cells, there is exhibited some form of learning ability. . . . In the simplest types, this ability consists in learning *not* to persist in certain behavior which would be in the normal instinctive reaction to a given situation. In more complicated animals, with nervous systems, the possibility of conditioned reflexes arises, where one stimulus can act as a sign of another one, provided the second follows within a short interval of time. . . .
>
> But when we ask whether animals can acquire sufficient knowledge of the processes of Nature to manipulate them for their own benefit, we find vanishingly little evidence, except in the case of the higher apes, and even they seem to have great difficulty, for instance in joining two sticks together to make a longer one. They can utilize signs (a bell means food), but they seem to have little or no power to develop symbols. Without symbols, abstract thought is, presumably, not possible. They seem to think in pictures which restricts them to merely "optical" solutions of mechanical problems.[2]

It is evident that the transfer of ideas by means of abstract symbols is a procedure restricted entirely to man. While the anthropoid apes exhibit a behavior that has much in common with human beings, "including a marked imitativeness, they show no signs whatever of speech, nor learn in the slightest degree to imitate human vocal sounds." Experiments have been done with young apes in close association with human beings over long periods, but no evidence of imitation of language has been observed. Animals can use signs, but they show no evidence of unaided use of symbols. Therefore the origin of communication involving conveyance of ideas with symbols must have occurred since the emergence of man as a distinct species, certainly within the last few million years, which is scarcely more than a fractional per cent of the total of biological history.

The unique ability of man to communicate by means of abstract symbols has not always been recognized. In his fascinating book "Science, Its Method and Philosophy," Brown points out that Darwin recognized no fundamental difference between man and animals with respect to speech. But Darwin did not differentiate between signs and symbols. Brown probably understates the case somewhat when he comments:

> The distinction between an animal that can make use of speech and one that cannot, remains one of the greatest importance.[3]

The great British biologist J. Z. Young says:

> Man's large brain is used to develop an intricate social system, *based mainly on communication by words.*[4]

The late social anthropologist Clyde Kluckhohn remarks:

> Nothing is more human than speech of an individual or of a folk. Human speech, unlike the cry of an animal, does not occur as a mere element of a larger response. Only the human animal can communicate abstract ideas and *converse about conditions that are contrary to fact.*[5]

Sophocles understood the relation of abstract thought, described by symbols, to the existence of human society.

> Of all the wonders, none is more wonderful than man who has learned the art of speech, of windswift thought and of living in neighborliness.[6]

In reviewing the higher attributes of man, Young points out:

> The factor we have been ignoring is that these special features of man are all due to the fact that he has developed far beyond other animals in the power of communication between individuals. Biologists have so far neglected to give full attention to the significance of communication in our species. The subject has been forced on their attention in recent years by the great development of mechanical aids to communication. There has already been some useful cooperation between biologists studying the brain and the engineers and mathematicians responsible for radio, television, and similar new devices. One result has been the comparison of the brain with calculating machines, but that is really only a detail. What is much more important is that we are now beginning to understand the importance of communication itself as a human activity. By thinking about this we shall find, I believe, a remarkable clarification of our ideas. . . . Proper use of communication has been the chief secret of the success of human societies in the past, and . . . it will certainly be so in both the immediate and the more distant future. Evidence of this is the fact that very extensive parts

of the brain are concerned with speech. We are only beginning to understand, however, how the brain works to produce particular methods of speaking. Societies certainly change their methods of communication through the centuries. Recently the Western world has developed a whole set of new techniques for the transfer of information. As a result, cooperation between individuals has improved, and better and better tools and machines have been produced. Men have gradually learned the great advantages that come from being able to convey information fully and exactly to each other.[7]

Considering our international troubles at the moment, Young's conclusion sounds a little Utopian. Nevertheless, our failures may well come from our inability to use properly the fundamental means for communication. Efficient communication through abstract symbols is at the very root of the civilizing process.

Not all languages are equally efficient in conveying ideas; most languages, to a varying extent, lack precision, and the development of an objective and precise science in the Latinic languages is probably no accident. As Clyde Kluckhohn has emphasized, "A language is a philosophy," and different languages invite attention to, and involve images of, quite different aspects of our general environment. To "think" in a language requires that we participate in the philosophy of its people. This is why communication between diverse peoples becomes so difficult and why translation of their languages is sometimes meaningless. A study of the origin and history of language is therefore essential to a deep understanding of communications.

2

The precision of communications, that is, the exactness with which the original image is correctly reproduced in the mind of the recipient, involves a host of intervening factors.

1. The capability of the intervening symbology—words, music, pictures, actions—to express the initial image in the required detail.

2. The capability of the originator to phrase, and of the recipient to interpret, the appropriate implications of the symbols employed, that is, mutual familiarity with the symbols.

3. The rate of transmission with respect to the available channel capacity for conveyance of the properly encoded symbols. Irrelevant signals, i.e., "noise," reduce the channel capacity by increasing the redundancy required to convey the necessary image with a defined measure of precision.

Until very recent times, indeed until the advent of information theory, men were inclined to deal with communications as the empirical conveyance of something called a "signal" from one place to another. A message was given to communication engineers, whose task it was to deliver it to its destination in one way or another as quickly, and with as little distortion, as possible. Ideas concerning efficiency of encoding and transmission with definable precision were only vaguely sensed, so that grossly inefficient, brute-force methods were necessary.

In this narrow sense, the development of the mechanical transfer of messages doubtless begins with the messenger who was employed in war in ancient times, or who traveled the earliest trade routes, carrying news of events. Evidence of early messengers goes back perhaps a million years, when news of a new weapon, a better axe or an improved spear, traveled from one group of men to another. We can trace the movements of such messages by their consequences on the lives of early man, whose recognizable traces show the social evolution of one group relative to another.

As man developed in war and in commerce, other mechanical means of communication came into use. From very early ages, sound signals from drums and smoke signals were used either casually or in organized nets to transmit information. Communications in these early beginnings rarely went so far as to spell out a message in detail; rather they were concerned with transmission of a few crude, prearranged, and generalized symbols, incapable of expressing shades of meaning.

Improvements in the mechanical modes of communication have generally occurred as a result of social pressure from war, from commerce, and from political development. The early Phoenicians signaled by movement of a sail. Slowly, methods of signaling with flags, guns, and other devices came into use. There is a long history of the improvement of flag signaling for military applications through the use of different flags, of the same flag in different positions, and of pairs of flags in different positions. For transmission in real time, such methods were restricted to a limited number of prearranged messages, since abstract encoding of symbols had not yet been invented.

For exact transmission, the messenger remained in vogue. Alexander, Hannibal, and Caesar all possessed an elaborate system of relaying by which messages were carried from one post to another by mounted messengers traveling at top speed. We know from "The Book of Marco Polo" of the fabulous message system across Asia established by the Genghis Khan at the close of the twelfth century. He not only created messenger posts from Europe to Peking, but also utilized that extraordinary messenger, the homing pigeon. As he advanced upon his conquests, he established pigeon relay posts across Asia and Eastern Europe to provide 70-mile-an-hour messengers to carry instructions from his capital to the provinces of his far-flung empire. It is said that, under the Khans, official communication from Europe to Asia required only a week.

With the words "What hath God wrought," transmitted by the electric telegraph of Samuel F. B. Morse between Washington and Baltimore in 1844, a revolution was triggered, not only in fast mechanical conveyance of information, but in the social organization of the world's peoples. Not only did the telegraph itself provide a completely revolutionary means of real-time, long-distance signal communication, but the development of the Morse code of dots and dashes was adapted to other kinds of signaling. The invention of the Morse code introduced a new element of flexibility and accuracy into signal systems, with surprisingly efficient encoding of abstract symbols that could be transmitted in any desired combination. It is the forerunner of our present binary "yes-no" encoding.

The Morse code is a highly formalized and abstract method of writing and thinking. As used at present, it involves only the translation of some letters of the alphabet and some words of a language into sharply defined "yes-no" signals that combine to express the letters and the words. But as any experienced telegraph operator knows, one does not read Morse letter by letter, but in words or in sentences, or even in ideas. It is a different mode of expression of the language *and with sufficient familiarity can come to represent the language itself without the intermediate step of formal translation through letters or sounds of words.* A proficient operator can not only carry on a free conversation in Morse code, but also express simple emotions by the intonation of his transmission, by ever so slight variations of spacing, the "swing" of his key!

When I was in the Antarctic in the late 1920s, several of the expedition members carried out, among other duties, radio-telegraph operations for many hours each day. Much of this operation was done with the excellent operators of the *New York Times* at a speed in the neighborhood of 45 words per minute, not much less than ordinary conversation. Because of the sparseness of population in our locality, the largest portion of our daily conversation was necessarily carried on in Morse code. One morning, when I was called for watch, I was startled out of a sound sleep to realize that I had been dreaming in Morse code!

In these days of information theory, this story has a moral. Perhaps we are approaching the stage at which we can dissect and encode the images of the mind with the ultimate efficiency and precision (the information theorist would say "approaching the entropy of an idea"), avoiding the encumbrance of an inefficient, intervening language, to which we are attached by chance of birth. The power of mathematical symbology points strongly in this direction.

In 1874, the telephone of Alexander Graham Bell initiated real-time transmission of speech by electrical encoding and replication of sound. The discovery and employment of electromagnetic waves, from Maxwell and Hertz to Marconi and the present, have come in the past three-quarters of a century. Encoding, transmission, and replication of visual scenes at a distance and in real time, i.e., television and picture transmission, have emerged from the contributions of many minds, as one after another natural phenomenon has been discovered, explored, and brought under man's control. But the story is far from its end, for in discovering and exploring each phenomenon, we have uncovered a multitude of new paths to follow.

Today, communication enters our lives and shapes our society in so many ways that it is easy to overlook the multitude of its facets. Aside from the telephone at our hand and radio and television in the living room, our newspapers are assembled by rapid communications from every part of the earth. Communication provides the senses of ships at sea, aircraft in flight, and rockets in space. Neither the aircraft industry nor space flight would be possible without the variety of today's communications.

Fast communications inform the weather forecaster, in nearly real time, of conditions seen by a multitude of observers. As late as 1926, the loss of life in hurricanes in the United States was 161 per 10 million dollars damage. Today, fewer than two human beings are lost in hurricanes per unit of damage. In 1900, more than six thousand were killed in the Galveston hurricane; during the more severe hurricane Carla in the same area this year, the loss of life was less than a dozen. The cooperation of weather stations, ships at sea, aircraft, and people everywhere, through organized networks of communication, for data and for warning, has greatly reduced the danger of storm. With meteorological satellites, coupled to even more complex circuit networks and data links of very large capacity, we can anticipate further improvement.

The defense activities of our nation employ fast communications in many roles, as a critical part of their operation. The development of defense systems has been a keystone in the evolution of new communications methods and devices.

Methods of transmission have also multiplied (Figure 1). The voice may travel by wire or through a wide-band cable; it may go by direct radio or be reflected from the ionosphere; it may be propagated by tropospheric or ionospheric scatter, or perhaps in the future by scatter from dipoles in space. Certainly, within the next few years it will be relayed by space satellites, either at low levels encircling the earth or hovering over our heads in a 24-hour orbit. Each of these methods of propagation involves a highly specialized and voluminous technology, painfully discovered, tested, and developed over the years. Yet, new methods of propagation are at hand. With the discovery of coherent light radiation from the optical maser we have found, in the nick of time, a solution to interspace communications across distances of millions of miles.

In looking back over the last century of communications, one is impressed by the strong interaction between communications engineering and pure physics. On the one hand, pure physics has provided concepts and devices for better communications. But on the other hand, numerous advances in physics have been predicated on the technological dexterity and on the clues and insights produced by the practice of communications.

In the field of transmission, Maxwell's theory of electromagnetism provided a basis for further research. But in the practice of communications, the unexpected phenomenon of long-distance short-wave propagation, encountered about 1918, seemed to run contrary to this theoretical foundation. Out of the natural processes thus discovered there

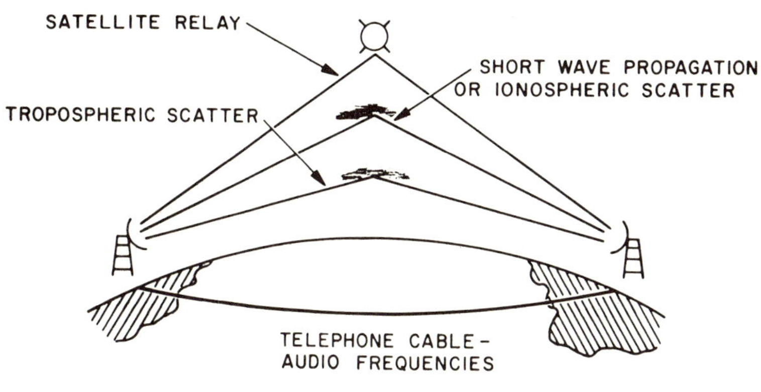

FREQUENCY		MAXIMUM RANGE
1000 - 10,000 Mc	SATELLITE RELAY	1/2 CIRCUMFERENCE
3 - 30 Mc	SHORT WAVE PROPAGATION	1/2 CIRCUMFERENCE
30 - 100 Mc	IONOSPHERIC SCATTER	1200 MILES
300 - 1000 Mc	TROPOSPHERIC SCATTER	800 MILES

Figure 1. Methods of signal propagation.

emerged an extensive and radically revised physics of the outer atmosphere and of solar-terrestrial relations. Thus an important branch of physical knowledge arose in response to discoveries made in practice.

Through this interaction of the work of the theorist and the engineer, communications has been a fruitful collaborator in the process of physical detection and measurement of phenomena to which our senses are not directly responsive. Science would be sterile and superficial in the absence of the dexterity and the insights that a progressive technology provides.

The most powerful and pervasive development in communications in this century has been Claude Shannon's discovery and elaboration of information theory which dates from 1948. In his first telegraph, Morse realized that "open wires" were necessary for high-speed transmission of Morse code. As early as 1855, Lord Kelvin had computed certain transmission properties of a submarine cable. Following the contributions of many others, Nyquist proposed, and later Hartley developed, the idea that the amount of information that can be transmitted over a perfect circuit is proportional to its bandwidth multiplied by the time of transmission. But how are we to measure the amount of information objectively, or how are we to specify the precision and actual rate of transmission in the presence of interfering noise?

Shannon showed in formal mathematical terms how to compute the entropy of an ergodic message source (Figure 2). Entropy, as the word is used in information theory, is the limit of efficiency in encoding a specified message expressed by the number of bits, i.e., yes-or-no signals, required per symbol. Thus the absolute minimum number of binary digits required to convey a message in a noiseless channel according to information theory is its entropy in bits per symbol times the number of symbols. The entropy of English comes out to be pretty close to one bit per letter. With this knowledge of the entropy of messages, mathematicians have been stimulated to develop methods of efficient encoding (such as the Huffman code, for instance) which approach the entropy of the message.

$$H = \sum_i P_i I_i = -\sum_i P_i \log_2 P_i$$

H = Entropy of the source generating the message

P_i = Probability of the i-th letter in the message

I_i = Information conveyed by the i-th letter

Figure 2. Entropy (average information) of a message source.

$$C_{bits/sec} = W \log_2 (1 + \frac{S}{N})$$

C = Channel capacity

W = Channel bandwidth

S = Signal power available

N = Power of interfering Gaussian noise

Figure 3. Capacity of continuous channel with additive noise only.

Information theory further shows that for the noisy channel (Figure 3), the channel capacity is gradually reduced by a theoretical amount of redundancy that must be added to the message to ensure error-free transmission. Then, as J. R. Pierce points out:

"The whole problem of efficient and error-free communications turns out to be that of removing the somewhat inefficient redundancy which they have, and then adding redundancy of the right sort in order to allow correction of errors made in transmission." [8]

From information theory, the limitations on encoding are known precisely by determination of the entropy of the source. For a given source, we can compute the maximum rate of error-free conveyance over the noisy channel in terms of its parameters.

The marked leadership of the United States in all aspects of communications theory and technology has resulted from the strong interaction of our industrial complex, our universities, and various government agencies. The Bell Telephone Laboratories, in which Nyquist, Hartley, and Shannon all produced their original work, provided a powerful stimulus to the development of fast, error-free communications. Among universities, the Massachusetts Institute of Technology has played a memorable role through the contributions of its special laboratories, as well as those of its president, Julius A. Stratton, of Norbert Wiener, and of many other individuals. Our successful experience in communications is an outstanding demonstration of the power of a free enterprise in producing national leadership.

3

The future of communications is inextricably entwined, both technically and in a social sense, with the growth of the new technology of which communication is so much a part. In 1922, Alexanderson predicted a maximum of 12 transatlantic radio channels. By 1960, through modern cable and radio, the actual number had grown to 136, an increase by a factor of about 10 in 40 years. We clearly foresee that in the next decade, with the coming of space communications, the long-distance channel potential will become 100 to 10,000 times as great. This is an increase of the order of 1,000 over Alexanderson's estimates, and of 100 over the present limited channel capacity. This radical improvement in the immediate future is symbolic of the current velocity of technological change.

Seen in perspective, the important revolution in technology began in the mid-twentieth century, and was symbolized by such advances as information theory and the transistor.

Until the beginning of this century, science was not a strong enough force to affect technology greatly. Engineering was still largely influenced by the empirical practices of the Babylonians, the Greeks, the Egyptians, and the Romans. True, the discoveries of Galileo and Newton showed how to mathematize empirical practices, and so to build bigger and better structures, but the fundamental concepts were the same. The discoveries of Franklin, Faraday, Henry, Maxwell, and Hertz provided a basis for the use of electricity and radio. But the technology of applying these discoveries drew heavily on earlier empirical engineering. The development of the molecular ideas of Democritus by Dalton, Priestley, and Lavoisier culminated in chemistry and chemical engineering. But even these were largely empirical, because of a lack of understanding of the forces that held matter together. "The crafts have often failed to learn from the theorists almost down to our day."

But a remarkable change came with the turn of the century. A series of scientific discoveries yielded powerful new insights into nature. With Einstein's theory of relativity, we found that atoms and electrons do not exist in the simple three-dimensional world to which we are accustomed, but in a four-dimensional space describable only by a tensor. With Rutherford, we learned how atoms are built. With Bohr, Schrödinger, and Dirac, we discovered that atoms and electrons do not obey the laws of ordinary mechanics, but the new laws of wave mechanics. In the evolution of information theory, Shannon used the concept and mathematics of hyperdimensional space to prove his classic theorems on the noisy circuit.

Relativity, wave mechanics, atomic structure, information theory—these are all examples of the revolutionary scientific discoveries of our century. These new concepts have given altogether new insights into nature. For the first time, science has acquired the depth to influence technology profoundly.

These discoveries are abstruse and relatively new, so that it has taken many years for a significant number of scientists and engineers to grasp them completely. Not until mid-century were these ideas understood and applied on a broad enough scale to bring about the profound effects on industry now observed. In the past ten years, nuclear energy, the transistor, space exploration, the high-speed computer, and data processing, among others of a wide variety of new industrial opportunities, have appeared in quick succession. It should be stressed that this is but the beginning. Our concepts of industrial processes and controls, of materials, structures, and communications are destined to change radically because of man's new powers to cause nature to do his bidding.

The new technology is drawn not so much from old empirical practices as from competence in theoretical science. Through fundamental understanding of natural phenomena, man has suddenly discovered that he can do completely new things and accomplish old tasks in a vastly more efficient manner. This is the context in which to estimate the future of communications.

The revolutionary character of the new technology is strikingly illustrated in methods of generating electric power. With the old practical technology, man was forced to adapt the ancient axle, turned by an inefficient engine, to rotate electric conductors in a magnetic field. By this method, he now achieves at best a conversion efficiency of heat to electricity limited to perhaps 35 per cent. With today's magnetohydrodynamics based on relativity and wave mechanics, we foresee direct conversion of heat to electricity without mechanically moving parts, in unlimited quantities, with efficiencies of 50, 60, or even 70 per cent. The efficiency and the vast range of functional applications of this new technology will displace still further the old "practical" technology in the decades ahead.

The processes of communications, and more broadly, of electronics, are ever more closely entwined in this new technology. Working in the four-dimensional world of electrons and atoms, electronics engineers are able to assemble materials into circuits of great reliability and complexity, with minuscule requirements for space and power and with broad applicability to all forms of signaling, computation, and control. These applications are now superseding earlier and clumsier methods of direction and control of industrial and commercial activities. Because of their economy and extreme flexibility, electronic methods are destined to become the nerve system of industry, entering into every aspect of the industrial process. Furthermore, electronics offers not mere replacement of crude mechanical methods of linkage and control, but concepts of a new order, which permit revised and fully functional industrial and commercial organization.

Coupled with forthcoming long-distance communications of great capability and reliability, made possible by instrumented earth satellites, these new applications present the prospect of widespread control of industrial complexes over large geographic areas. The leadership of this technological revolution must come from men who command the ideas and mathematics of the new science. This places new requirements on the content of engineering education, on specialized higher education, and on new professional methods of advancement, so that members of the electronics profession can keep pace with this revolution.

The United States communications industries involve a business effort variously estimated at between 10 and 20 billions of dollars, depending on what is included in the figures. This business is the vital key to the functioning of a much larger proportion of our total industrial and governmental activity. Because of the efficiency that communications can add to industrial functions, to government operations, to human convenience and happiness, it must take an ever-larger share of an ever-increasing gross product. With the limitation on transocean channel capacity removed by satellite communication, we can see that communications must expand explosively in the decades ahead.

The broader implications of the new knowledge in communications raises many questions. Can ideas be encoded and conveyed without the intervention of an inexact, encumbering language? How can ideas be subdivided into their most basic elements, and what is their entropy? How can ideas be encoded efficiently? These are problems for the linguist, the mathematician, and the psychologist, certainly invading the general field of symbolic logic.

> I myself think that information theory has provided psychologists with a new and important picture of the process of communication and with a new and important measure of the complexity of the task. It has been important in stirring them up. . . . Information theory provides a central, universal structure and organization for electrical communication, . . . [as yet] it constitutes only an attractive area in psychology. It also adds a few new and attractive expressions to the vocabulary of workers in other areas.[9]

4

The new vocabulary of communications offers the opportunity to reexamine some old problems. According to information theory, the channel capacity corresponds to the bandwidth only in a noiseless circuit. As noise increases, the channel capacity diminishes, for noise signals are mistaken more frequently for the desired signal, and more redundancy must be introduced. At the limit, all channel capacity disappears.

In conveying images between minds, the channel is always noisy. Noise can consist of misapprehension, fear, envy, hate, indeed any emotion that is stimulated by the signal. Noise can be generated by misconstruing a word or phrase, or not understanding it at all. One suspects that, in direct conversation or study, two minds are connected by a limiting bandwidth whose channel capacity is easily reduced or destroyed by the noisy interference of irrelevant emotions. It would prove interesting if quantitative expressions could be found for these basic processes of communications among men.

In brainwashing, i.e., excessive repetition of one idea, the individual seems to develop emotional resonances. Whatever message is sent, only one standard response is stimulated, far above the level of all others. Every message is interpreted in the light of this one idea. The ingrained communist, the religious fanatic, the superpatriot, all seem to have sharply tuned, but readily excited, resonances with very high "*Q*" that overpower the incoming message. Perhaps these emotional resonances and their origins can be described mathematically in terms of filtering practices.

The speed of modern communications tends to force all decisions into the pattern of tactical decisions. A tactical decision is one which must be made against a time deadline, as on a battlefield or in passing another car on the highway. Tactical decisions tend to employ motor responses of the brain, and therefore encourage the response of emotional resonances. High-speed communication tends to deemphasize strategic responses, that must be developed as a result of long research, study, and carefully reasoned conclusions. The more important strategic decisions are perhaps better geared to the speed of the sailing ship than to the instantaneous response required by the telephone or the cable. Hence, in light of today's communications, we need some quantitative guide to the character and quality of decision making.

The psychologist and the linguist are only at the beginning of the new task that the communications engineer has set for them. Perhaps they will find that, in our present ignorance, war is inevitable and quite beyond our control, because, though nations speak, they cannot communicate through human noise levels which narrow the intervening channel capacity. Perhaps we must find reliable symbols whose entropy and method of encoding permit their conveyance through channels of negligibly low capacity. Perhaps this is the meaning of Reinhold Niebuhr's penetrating comment that rule of law is not possible until peoples are bound together by threads of common cultural understanding. Perhaps we must find symbols which distinguish sharply between the tactical and the strategic response.

Communication, a major element in lifting man from the brutality of the past, is at the root of civilized progress and man's ability to control his environment and organize his social structure. Unquestionably, improved communications have increased man's productivity, his safety, his welfare, and his happiness. Perhaps, too, our growing knowledge of the communication process will eventually permit communication among peoples in the fuller sense of conveying images more precisely, or at least without damaging degradation, so that the alternatives can be evaluated and chosen more precisely and wisely. As knowledge is created, our access to the unknown is enlarged.

"Comparing something unknown with something already known makes it possible to talk about the unknown." [10]

REFERENCES

1. G. B. Brown, "Science, Its Method and Its Philosophy," p. 30, Norton, New York, 1950.
2. Ref. 1, pp. 29, 30.
3. Ref. 1, p. 32.
4. J. Z. Young, "Doubt and Certainty in Science," p. 5, Oxford, New York, 1951.
5. Clyde Kluckhohn, "Mirror for Man," p. 145, McGraw-Hill, New York, 1949.
6. Antigone.
7. Ref. 4, p. 6.
8. J. R. Pierce, "Symbols, Signals, and Noise," p. 164, Harper, New York, 1961.
9. Ref. 8, p. 249.
10. Ref. 4, p. 31.

Ivan A. Getting, President, Aerospace Corporation, El Segundo, California, holds the President's Medal of Merit for his wartime contributions to radar. He has been a member of many government committees and special study groups concerned with national defense.

CHAPTER 4

RADAR

1

The story of radar is also the story of our times, of political turmoil, a revolution in scientific methods and management, and an explosion in applied science and technology. The menace of Nazism and fascism brought radar into being, and the continued threat of communism has spurred the growth of its capabilities to the present high pinnacle. The development of radar in England and in the United States has left in its wake new concepts of organized research and development, involving the integrated efforts of large teams of scientists, as at the Lincoln Laboratory. The evolution of radar, possibly more than that of any other technology, has manifestly influenced our lives and those of the next few generations, by spawning allied technologies and supporting them through techniques, components, and concepts which arose from the primary line of radar development.

Microwave technology resulting from radar has found application in microwave communications links, now spanning continents; and higher-power tubes have made possible scatter communications linking even remote areas like Greenland and Alaska. Wide-band amplifiers and pulse circuitry, the handmaidens of radar, together with improved radar oscilloscope techniques, presaged the widespread availability of television. Because radar has given man a sensing capability in many ways far superior to his common senses of sight and sound, the commensurate ancillary equipment required has forced the development of servomechanisms, analog computers, and digital computers, which have themselves gained much of the requisite technology from radar technology, to bring about what is clearly a revolution in man's utilization of data and information. Finally, radar has paid its debt to science itself by providing tools, as in microwave spectroscopy and radio astronomy, which have materially advanced our fundamental knowledge of matter and of the universe.

2

Before embarking upon the history of radar, I wish to draw attention to the invention of the carbon-filament electric light bulb by Thomas Alva Edison in 1879. This, from a case-history viewpoint, was a nice invention. The need was clear; the invention filled the need. The economics of the situation are also tractable; one can compare the relative cost of so many lumens as produced by gas mantles, for example, and by electric light bulbs. In the minds of many, the inventor is clearly identifiable. Nevertheless, the results, though tremendous, did not add a new capability to man.

In contrast, the development of radar was quite different. In large part, the difference arises from the fact that the capabilities of radar go far beyond the expectation of the early workers, and these capabilities do not merely replace other methods at lower costs, but add significant new dimensions. The early work in radar, in particular, was a search for a solution to a hitherto intractable problem: the detection of aircraft at long distances, at night and during daylight, in rain and through clouds. Because the process was evolutionary in character, there was no single invention, and because the need was largely military in character, the work was shrouded in military security with purposefully introduced blocks in the exchange of ideas, not only between scientists of different countries, but also between workers in the same countries. There is thus no single inventor of radar, nor even any one country, uniquely deserving of the credit.

A. P. Rowe, in his record of a part of the British World War II radar effort, wrote:

> Those who do not move in scientific circles are inclined to think of inventors as men who, like Archimedes, spring from their baths shouting "Eureka" when some great truth is vouchsafed them. The public and therefore the press usually demand that the name of one man shall be heralded as the inventor; it is simpler for all concerned. Few important advances in applied science happen in this way and certainly radar did not.[1]

As a matter of fact, the development of radar was pursued independently during the 1930s in at least five countries: Great Britain, the United States, France, Germany, and Canada. Whereas the motivation in France, like that of Edison's work with the light bulb, was nonmilitary, the motivation in the other countries was essentially, though not wholly, military. In the end, it was the military need that dominated, and certainly, up to the start of World War II, only the military effort produced lasting contributions to the technology of radar.

Henry Guerlac, in an excellent history of the antecedents of radar, has said:

Each striking instance of parallel and independent discovery raises a number of fundamental historical questions. When, moreover, a real burgeoning takes place, we are led to ask what the conditions were that favored it. The main influences are obvious. First of all, when a serious effort is put behind a development, as in the case of radar, it is evident that this development satisfied a clear and urgent need. Second, since the principle of *nihil ex nihilo* applies quite as surely to the history of science as to science itself, it is obvious that some key principles or central ideas in pure science must have been the point of departure. Lastly, if success is finally attained, it is certain that the state of the art, that is to say the perfection of engineering skills in this and neighboring fields, must have reached a point where success was fairly well assured.[2]

It is not my intent to provide a comprehensive history of radar. As a matter of fact, no one has dared, much less produced, a comprehensive treatment of the subject, and I will not rush in where professional historians fear to tread. Instead I shall recount briefly the three basic influences earmarked by Guerlac: the need, the basic scientific principle or central idea, and the availability of the requisite engineering skills.

Clearly, the dominating need which cradled the development of radar was military in character. The airplane was introduced into the arsenal of weapons in World War I, a fledgling playing a very small role. But by the early 1930s, it was clear to most military thinkers that the speed and logistic capability of this weapon system might have a decisive effect on any new, but then still inconceivable, war. As any new threat evokes a reaction, so the airplane threat stimulated efforts to find a defense. But in the period 1930–1935, the methods, suggested and attempted, for meeting the new danger by passive infrared detection, passive radio tracking, and sound tracking, all proved ineffectual in providing adequate data for defense against attacking aircraft.

By 1935, the growth of Nazism and fascism, clearly aiming at world domination through the force of arms, raised the urgency of the need for adequate defense to the point where, in England and in the United States, the highest elements of government became involved. The effort in England was perhaps the more dramatic. The occupation of the Ruhr in March 1936, the occupation of Austria in March 1938, then the shameful meetings of Berchtesgaden, Godesberg, and Munich in September 1938, leading in rapid crescendo to the occupation of Czechoslovakia in the spring of 1939, left no real assurance of "peace in our time." England was within range of the bomber fleets of the rapidly growing Luftwaffe. Churchill, who since 1933 had been warning all who would listen of Nazi aggression and the German military build-up, became more and more influential.

In the short period from early 1935, when Sir Robert Watson-Watt made his first proposal to the British Government for a pulsed radar to operate at 25 Mcps, to September 30, 1938, the date of the infamous Munich Pact, radar in England went from concept to 24-hr operational use. When war broke, 700 British fighters faced 2,000 Nazi bombers. With the aid of the so-called CHAIN radar stations (Figure 1), the battle was joined. Without them, England would have been razed, as were Poland and the Low Countries. Tracking the German planes as they assembled over continental airfields, these radars, magnificently exploited, placed the limited number of British fighters in attack positions. Radar reversed the traditional advantage of surprise normally

Figure 1. Tx towers, CHAIN radar. (Reproduced by permission of Her Britannic Majesty's Stationery Office. Crown copyright reserved.)

available to the attacker, and permitted instead the concentration of defending forces at the critical points and at the right time. The Battle of Britain, in the summer of 1940, could not have been won without radar, and the world will never forget this first rebuff to Hitler's nefarious schemes or Churchill's famous statement in referring to the pilots of the Royal Fighter Command: "Never in the field of human conflict was so much owed by so many to so few."

In the United States, the need in the early 1930s was not as apparently urgent. Our continental isolation and the limited range of aircraft did not present a national threat. Nevertheless, much credit is due to the personnel of the Naval Research Laboratory and to the Signal Corps Laboratories. Their pioneering work, which preceded that of the British, resulted in the development of operational radars at higher frequencies and, consistent with our different military needs, designed for mobile war. Thus the United States Navy installed its first production operational radar aboard ship in December 1940, and the Signal Corps had its early warning radars in operation during the attack on Pearl Harbor, December 7, 1941.

The need for improved radar performance was accentuated by the shift in the Battle of Britain from mass day raids to night bombing and, later, by the increase of German submarine warfare, which threatened to end the European war by cutting the life line of supply from the United States. Under these stresses, the radar development efforts in England and in the United States were united. In September 1940, a British mission led by Tizard visited the United States. The visit was coincident with the establishment by President Roosevelt of the National Defense Research Committee under Vannevar Bush. K. T. Compton and Alfred Loomis, among others of the National Defense Research Committee, met with the members of the Tizard mission. Directly following from these activities was the establishment here at M.I.T. in October 1940 of the Radiation Laboratory, whose function, by agreement with the Army and the Navy, was to push the art of radar to higher, or microwave, frequencies, while the armed services, splendidly successful with equipment at wavelengths of the order of a meter and longer, were to be engaged in the stupendous tasks of further improvement, procurement, training, and maintenance.

Thus the need for radar toward the end of the period 1930–1940 was very urgent. Let us now turn to Guerlac's second influence essential to burgeoning invention: the basic scientific principle. The radar of World War II, as is indeed true of most radars today, employed short pulses of high-frequency radio waves. These are transmitted through space, generally in a well-defined beam, at the velocity of light. Upon striking an object, some of the energy is reflected back toward the receiving antenna, where the received signal is amplified and processed. The direction of the target is crudely established by the directivity of the beam, and the range is determined by measuring the elapsed time between the transmitted pulse and the received echo pulse, ten-millionths of a second corresponding to about one mile. In fact, the word "RADAR" was coined by the United States Navy from *RA*dio *D*etecting *A*nd *R*anging.

Were the basic scientific principles understood in the 1930s? They were well established by the turn of the century. In 1873, James Clerk Maxwell in his "A Treatise on Electricity and Magnetism" defined the basic mathematical theory; and Heinrich Hertz in 1887–1888 demonstrated the existence of high-frequency radio waves. By 1895, Peter Lebedew had produced 6-mm waves, and many scientists had clearly demonstrated the propagation speed, polarization, reflection, and refraction of these waves. The development of radar was not, therefore, held up by the lack of basic understanding. Indeed, patents describing the technique of electromagnetic radar were issued as early as 1904 to Christian Hulsmeyer of Dusseldorf,[3] and pulsed techniques using oscillators were designed for echo depth sounding by Fessenden in Boston at about the same time. The missing link, then, was neither the need nor the basic scientific theory, but rather the third essential influence: the requisite engineering skills.

It is probably true that these requisite engineering skills have been the pacing element from the 1930s to this day. Therefore radar has been almost exclusively a field of applied science and engineering, and each step forward has led to new capabilities, opening up new avenues of application and new demands for ancillary equipment and additional development.

It is essential that we hold in proper perspective the electronic engineering accomplishments of the early thirties. The unique tool of electronics of that day was the triode amplifier oscillator. This vacuum tube, invented by Lee De Forest in 1907, made possible both oscillators of stability and reasonably high power and also detectors and amplifiers for use in receivers. But in the 1930s, radio communication was still in its first ten years of commercial use. Early work by Marconi and others had indicated that low frequencies were better adapted to long-range radio communication. The technology of low frequencies was easier and well within the scope of the triode-electron-tube technology.

This application of science evoked a new scientific question: How could one explain that these low-frequency waves are transmitted around the curvature of the earth in apparent contradiction of Maxwell's theory, which predicts that radio, like light, is propagated along straight lines? This led to the hypothesis of a layer, or several layers, of free electrons in the ionosphere, acting like mirrors and thereby trapping the radio waves in a relatively narrow duct. In the measurement of the heights of these layers by teams from the Carnegie Foundation and by others, techniques similar to pulse radar were developed. Simultaneously, the employment by amateurs of the high-frequency bands, thought to be generally useless for commercial long-range application, resulted in productive techniques pushing the triode to ever-higher frequencies.

All these events formed the backdrop of the birth of radar in the 1930s: (1) transmitting triodes were available for use in the frequency region up to wavelengths of about 1 meter; (2) substantially more power, measured in terms of kilowatts, was available at lower frequencies in the neighborhood of 100 meters; (3) receivers were available in these frequency ranges, although at the 1-meter high-frequency end, the performance was poor.

To give in a short time a coherent and accurate account of the technological background of electronics as it pertained to radar in the thirties is impossible. It is, however, necessary to draw attention to three major components: the oscilloscope tube, the klystron, and the magnetron.

The oscilloscope (Figure 2) is best known to the general public today in the form of the television tube. Its availability in the 1930s was an engineering prerequisite for radar. The oscilloscope tube made possible the visual presentation of pulses shorter than a millionth of a second. Its versatility is well demonstrated in a plan position indicator as used in radar. First invented in 1897 by Braun, the oscilloscope tube, or cathode-ray tube, as it was then called, was much improved by A. R. B. Wehnelt by the introduction of an oxide-coated cathode. By the late 1930s, the oscilloscope was essentially a shelf item, but like all other electronic components, it was greatly improved under the exacting needs of World War II.

The klystron differs from the conventional triode tube in that it makes use of the finite velocity of electrons between the cathode and anode instead of being subject to the so-called transit-time limitation of conventional tubes. In the klystron, the electron stream from the cathode is first focused into a beam, and the velocity of the electron beam is then time-sequentially varied or modulated, so that as the electrons move down the tube, the faster ones catch up to the slower ones, producing bunching. These narrow bunches of electrons then pass through a resonant cavity, or rhumbatron, giving up their energy to radio oscillations of the cavity, much like a hammer striking, at appropriate intervals, a massive pendulum swinging by. By feeding back some of the energy from the output cavity to the modulating cavity, the klystron is caused to oscillate.

The principle of velocity modulation was first promulgated by A. and O. Heil in Germany. By 1939, the brothers R. H. Varian and S. A. Varian, in the United States, had developed excellent klystron oscillators. These tubes were used almost exclusively in the higher end of the frequency band as local oscillators for superheterodyne receivers. One of their chief attractive features is their susceptibility to mathematical analysis, and therefore to straightforward design. The klystron is today one of the principal sources of high power at microwave frequency. But in 1940, its useful power was of the order of 1 watt, much too low to permit a breakthrough as a power transmitter for effective microwave radar.

Figure 2. Oscilloscope tubes, 1940–1945.

The breakthrough, and the missing element in the development of microwave radar, was the multicavity magnetron. To those brought up in postwar electronics technology, the magnetron may not appear to be a remarkable device. On the other hand, to those who cut their teeth in the 1930s, it was truly amazing. Everyone then knew that high-power, high-frequency tubes were limited by cathode emission, and that whereas oxide-coated cathodes were the most effusive emitters, they could not be used in the presence of the high anode voltages required by high power. Everyone also knew that high-power tubes became very hot and that therefore they could be made only of heat-resistant glass and must be thoroughly outgassed. By the same token, the use of metals for electrodes had to be limited to tantalum, nickel, and tungsten, metals that could be outgassed in a vacuum at high temperatures by heating them cherry-hot through induced currents. Hence, how amazing that the high-power magnetron should have been made with an oxide cathode, with a copper electrical circuit built into the vacuum envelope, and that the container itself should be of copper, instead of glass!

The multicavity magnetron was first built in England in 1939 by H. K. H. Boot, J. T. Randall, and their associates, as a direct result of the pressure of war. Randall was a physicist acquainted with the cyclotron particle accelerator invented by E. O. Lawrence. The danger of the night bombing of England pointed to the need, in addition to the CHAIN low-frequency ground radars, of airborne radar to be carried by the night fighter-interceptor. The ground stations could, in principle, vector the interceptor within a few miles of the attacker, but the actual destruction of the night bomber through firearms required a closing and an accurate pointing. This could be done only by a radar carried by the fighter plane. Requirements of small antenna size and accuracy adequate for the aiming of guns clearly indicated the shortest possible wavelength: at least 10 cm, or approximately 4 in. This necessity brought about the invention of the multicavity magnetron.

Whereas the klystron lends itself to clear analysis, the cavity magnetron is a pernicious beast. While analysis is of some help as a guide, so that the principles are understandable, the details are at best messy. Nevertheless, the miracle did happen: the unorthodox tube did oscillate at frequencies of 3,000 Mcps, or 10 cm wavelength, and at a high power of the order of 20 kw peak power, a jump at these frequencies of about four orders of magnitude. When an increase in technological capability of one order of magnitude becomes available, marked advances can be expected, but with a sudden increase in capability of four orders of magnitude, an explosive technological revolution is possible.

The importance of this invention was immediately realized by the British Government, and when Tizard arrived in the United States in September 1940 on his unprecedented mission to establish mutual scientific help, he came bearing with him a small box containing the secret tube.

A. P. Rowe has underlined its significance:

> I suppose, however, that few in a position to judge would hesitate to name the cavity magnetron as having had a more decisive effect on the outcome of the war than any other single scientific device evolved during the war. It was of far more importance than the atomic bomb, which had no effect at all on the outcome of the German war and contributed rather to the shortening of the Japanese war than to its result.[4]

This new tube had an oxide-coated cathode in a copper container, when everybody who was anybody knew that pure metallic cathodes had to be used and when the experts knew that hard glass envelopes and special metals were essential. The lesson here reads "loud and clear," as they say in radio jargon. The thirties had seen the development of a radio technology based on voice communications, which required continuous transmissions. Hence the technology which developed—much of it by trial and error—was characteristic of that application. Pulsed radar, however, had basically different engineering requirements. While the peak power required for useful range was very high, of the order of the highest low-frequency radio stations, the pulses were very short, of the order of a millionth of a second, and the average power, which determines heating, was less by a factor of 1,000. Many breakdown phenomena are statistical in nature, and if the time interval is sufficiently small, like a millionth of a second, the chances are high that the breakdown will not take place. Hence the builders of the cavity magnetron were successful with designs wholly unacceptable to the prevailing technology. I doubt that the physicists-turned-inventors necessarily reasoned this out; I rather suspect they were unencumbered by too much information. The moral is, when facing a new situation, always go back to first principles.

Another consequence of the breakthrough was associated with the duty cycle. While the peak powers were high, the average power was relatively low. Since any electronic gear consists of many components, it is clear that these other components also were affected by the change. New classes of transformers, capacitors, wire, connectors, and test equipment were required. And whereas voice communication required a bandwidth of a few kilocycles, the short pulses of radar, with the sharp rise times and sharp tails needed for high range resolution, required a bandwidth of several megacycles, an increase of three magnitudes. It can be said that components technology of the 1930s was dominated largely by the generation of radio hams of the 1920s. But World War II, primarily through the growth of radar technology, brought about a sophistication, so that the increase in variety and versatility of electronic components subsequent to the war represents the highest application of the physical sciences to technology.

A detailed history of the progress of radar technology during the war and its effects from campaign to campaign has never been published. A portion of the scientific work done during the war years was described in the now famous M.I.T. Radiation Laboratory Series of 28 volumes. Much new material which was classified at that time has not been published.

It will have to suffice here to mention only a few of the battles and campaigns in which radar played a decisive role. There was first the Battle of Britain in the summer of 1940, when the flood of Nazi Luftwaffe daylight bombing raids was stopped. The night bombing of Germany by the Royal Air Force and the daylight bombing of Germany by the United States 8th Air Force, through the incredibly poor weather of that part of the world, were made possible largely by microwave radar: the H2S and the closely allied techniques of air navigation called Gee, Oboe, and Shoran.

Containment of the U-boat threat, particularly in the seas around Great Britain and in the Bay of Biscay, was a technical battle of radar wits. As the German submarines surfaced at night to charge batteries, they were attacked at first by planes equipped with British 1½-meter airborne radar. The Germans installed intercept receivers to warn the submarine skippers of the approaching planes. The Allies responded with newly developed radar equipment in the S-band, or 10-cm, region. The Germans in time, but not before suffering damage, installed the appropriate intercept receivers. As the state of the art, both technical and tactical, improved, each side introduced innovations. The Allies installed attenuators into the transmitters, which took advantage of the inverse fourth power law for reflected radar signals and the inverse square law for direct intercept. Thus, as soon as a submarine was spotted on the radar, the plane turned toward the submarine and the radar operator decreased the transmitted power at such a rate as to cause the intercept submarine operator to interpret the signal as that of a receding plane.

At about this time, too, the principle of radar flooding was introduced, so that a submarine intercept receiver picked up signals at all times. Since the submarine commander could not keep his craft submerged indefinitely, he had to take chances. Signals were always present, and an apparently receding airplane might or might not be a safe risk: under the circumstances, he was permitted only one mistake. The result of these radar tactics was the invention by the Germans of the snorkel, which reduced to a minimum the area available for reflecting radar signals. This innovation came too late to save the submarine. The Battle of the Atlantic had been won.

New techniques continued to come out of our scientific laboratories. The Normandy landings were monitored by the MEW radar built at the M.I.T. Radiation Laboratory (Figure 3). This, the highest-power (10 cm) radar of World War II, became the backstay of tactical air forces. The wartime operational control center of the MEW radar is shown in Figure 4. Figure 5 is a photograph of the plan position indicator of a MEW radar installed on the southeast coast of England: the date, 6 June 1944; the time, 0653 hours; the event, invasion of Europe.

Figure 3. AN/CPS-1(MEW) World War II microwave warning radar.

Figure 4. World War II operational control center (AN/CPS-1).

The coasts of England and Normandy are superimposed on the scope, which shows clearly the large traffic of Allied air power over the English Channel.

This equipment, with modified SCR-584 automatic tracking radars, not only gave our 8th and 9th Tactical Air Forces warning of enemy planes, for which they had been originally designed, but also made possible effective employment of our offensive air power. By providing accurate traffic control and by giving the tactical commander on the ground continuous information of his airborne planes, preferably equipped with the so-called rosebud radar beacons, it was possible to carry out accurate aerial reconnaissance and bombing, the latter often under conditions of poor visibility, as during the Battle of the Bulge.

Antiaircraft radar played a dramatic role in the Anzio beach landings (Figure 6), where our troops were concentrated in a small landing area. Nightly bombing raids by the Luftwaffe hindered the build-up necessary for a major offensive. The landing of a few antiaircraft batteries equipped with microwave automatic tracking radars in one night stopped the raids.

Toward the end of the European war, Hitler hurled first the V-1, and then the V-2, weapon against the population of greater London. The V-1 drones, available in mass production, flew at low altitude and at speeds greater than any fighter except a few Spitfires and Meteor Jets.

Each V-1 drone carried a warhead of 1,000 lb. In a few weeks, this threat was contained by moving several hundred microwave-radar-equipped gun batteries from around London to the coast of Britain. With a little practice, 95 per cent of the V-1s were destroyed while still in flight.

The V-2s, however, presented a different problem, one which we are now facing in the "ultimate weapon," the nuclear warhead ballistic missile. From the military viewpoint, the TNT-loaded V-2 appears to have been a poor weapon. It did some damage to London and much damage, although not critical to the war effort, to Antwerp. Several tracking radars were deployed in Holland. These tracked the missiles in flight and, by establishing the trajectory, computed the launching position. This enabled the dispatch of fighter-bombers to the camouflaged launch sites. This exchange between ballistic missile and radar defense was important, not so much because of its strategic significance in the European war, as because of the fact that it was the first employment of long-range ballistic missiles and an attempted counteraction.

The war in the Pacific was quite different from the European war in many respects, including the role of radar. The B-29 bombers, which replaced the B-17s and the B-24s, were equipped with second-generation microwave equipment, properly integrated into the airplane. The Japanese were not so advanced in radar-controlled antiaircraft as were

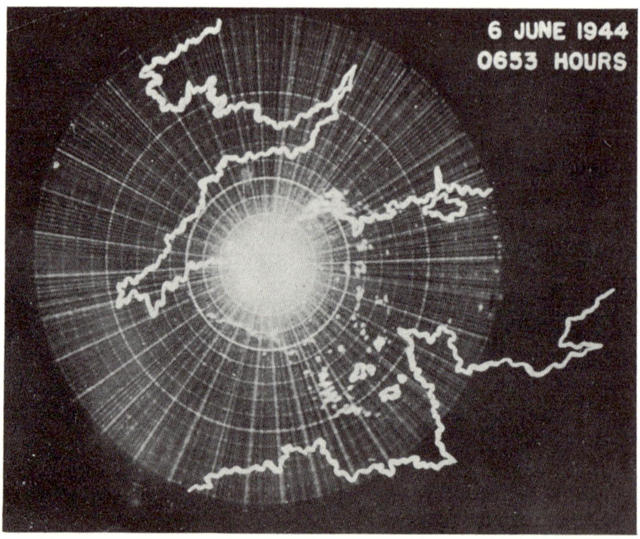

Figure 5. Invasion of France as seen by radar, 6 June 1944.

Figure 6. SCR-584, microwave antiaircraft fire-control radar.

the Germans, with their Wurzburg sets. Nor did the Japanese have the electronic jamming capabilities of the Germans. The effectiveness of United States radar-controlled B-29 bombing was great indeed, but from a radar point of view, one might say that it had become "routine."

At sea, the radar battle was also very effective. For the first time in the war, the Navy had an opportunity to carry out its primary function of destroying enemy seapower. The Navy's most important sets were the SG microwave family of surface 10-cm search radars, the SC meter-wavelength aircraft warning sets developed before the war, and the various fire-control sets, such as the FC and the FD operating at 50 cm. These sets gave the American Navy a tremendous advantage over the Japanese, who were often not equipped with any radar. In the Battles of Leyte Gulf, Surigao Strait, Samar, and Cape Engano, the Japanese Navy was destroyed, and the stage was set for the eventual defeat of the Japanese Empire.

It must be added, however, that in addition to its function in these dramatic events, radar played workhorse roles in many ways: station keeping in convoys, sea navigation, harbor control, air navigation, bomber training and assessment, weather forecasting, and in submarine attacks on surface ships.

When the war ended in 1945, the Office of Scientific Research and Development was disbanded, and large industries that had manufactured a billion dollars' worth of radar turned to the production of air-conditioning and television equipment. By 1948, the United States had virtually no air-defense capability, feeling secure in the possession of the atomic bomb and in overwhelming bomber superiority.

As a consequence, the low priority on ground radars and on all-weather fighters led to a general retrogression in equipment and systems. True, some advances were made, but not fast enough to keep up with the tremendous advances in long-range, high-speed bombers. For example, the Air Defense Command was not established until 1948, and the only available all-weather fighter was a converted two-seater training plane equipped with a hurriedly built radar copied from a World War II design.

Two events altered the situation: the first Russian atomic bomb, tested in 1949, and the outbreak of the Korean War in June 1950. The urgency of the problems of air defense resulted in a major effort in research and development, a large part of which was concentrated in newly established institutions, like the Lincoln Laboratory at M.I.T. The next decade saw the completion of the DEW (Distant Early Warning) line, BMEWS (Ballistic Missile Early Warning System), and SAGE (Semi-Automatic Ground Environment). A major command, NORAD (North American Air Defense), is now in operational control of all the air-defense resources of the United States and Canada.

3

In reviewing the substantial progress in radar technology during the past ten years (Figure 7), it is necessary to mention, if only briefly, the important contributions of masers, transistors, and other products of solid-state research. In addition to the tremendous gains resulting from semiconductors, the development and application of ferrites and ferroelectric ceramics have given microwave scientists new dimensions in design freedom, without which many of the present-day radars and bold new future ones would not be possible. Among the two most important classes of ferrite components are the nonreciprocal type, such as the isolator, which permits power to flow in one direction but not in the other, and the phase shifter and attenuator, which permit controlled cancellation of signals and the matching of tubes run in parallel to increase total power.

I shall touch on only four of the many areas in which radar technology advanced during the decade 1950–1960: the generation of high power, the development of more sensitive receivers, the reduction of ground clutter, and information processing.

1. HIGH POWER TRANSMITTERS	11. PULSED-DOPPLER RADAR
2. LOW NOISE RECEIVERS	12. CONTINUOUS WAVE RADAR
3. MILLIMETER-WAVE TECHNIQUES	13. MONOPULSE TRACKING
4. LIGHT RADAR (LASERS)	14. FREQUENCY SCANNING ANTENNAE
5. APPLICATION OF INFORMATION THEORY	15. LARGE TRACKING RADARS
6. PULSE COMPRESSION	16. SIDE-LOOKING RADARS
7. SLOWED-DOWN VIDEO	17. PHASE COHERENT MOVING ANTENNAE
8. FINE-GRAIN DATA TRANSMISSION	18. SIDE LOBE SUPPRESSION
9. STORAGE TUBES AND BRIGHT DISPLAYS	19. CIRCULAR POLARIZATION
10. MAP-MATCHING	20. FERRITE MATERIALS AND COMPONENTS

Figure 7. Areas of progress in radar technology, 1950–1960.

Other things being equal, the effective range performance of a radar set depends upon the power transmitted. To double the range requires sixteen times the power. There are other ways of increasing range: one can increase the size of the antenna or reflecting dish; one can increase the sensitivity of the receiver; one can employ some sort of time integration of the received data. Most of these approaches toward increasing range degrade the performance in other aspects. Increasing the size of dish narrows the beam and reduces the scanning rate of the system; the integration of received data slows down the data rate, reduces accuracy, involves equipment complexity, or all these to some extent. Then there are limits to how far the sensitivity of a practical receiver should be increased. All radars operate in a real world, where nature and man have created a background of ambient radio noise. In the end, for a given solid angle and a specified data rate, the average transmitted power is practically the only open-ended variable left. Thus in a review of where we stand, a consideration of the increase in power generation is basic.

At the end of World War II, the only useful generator of high-power microwaves was the magnetron. Its frequency range, expressed in wavelength, was from about 30 to 3 cm. Peak powers ranged from about 1 megawatt at the longer wavelength to about 30 kw at the shorter wavelengths. The average power varied from about a kilowatt to a fifth of a kilowatt. With these tubes in appropriate equipment, detection of large bombers at ranges of about 200 miles was feasible. Since then, several new types of tubes have been developed. Like the magnetron, they all depend upon the interaction of free electrons moving in a vacuum with an efficient radio-frequency metallic circuit. In general, most of the tubes can be placed in one of two classes, which are distinguished by the way in which the primary energy is converted into radio-frequency energy.

In one class, the primary applied electrical energy is converted into the kinetic energy of a beam of electrons. The beam may be contained or focused by a longitudinal magnetic field. The electric fields of the moving electrons impart energy to a radio-frequency circuit. In the klystron (Figure 8*a*), the first cavity is used to velocity-modulate the beam and the final one is used to extract the microwave energy, the coupling between the cavities, if any, being done externally. In the traveling-wave tube (Figure 8*b*), the reiteration of the circuit is continual and each repeated element is internally coupled.

In general, high power is obtained in these tubes because of large cathodes, which make possible high total emission; magnetic focusing, which allows high current densities in the limited interaction space; and the location of the heat-dissipating collector essentially out of the interaction space. Practical high-power klystrons (Figures 9, 10) are characterized by very high gain, as much as 60 db, but they are generally relatively narrow-band and require extremely high voltages.

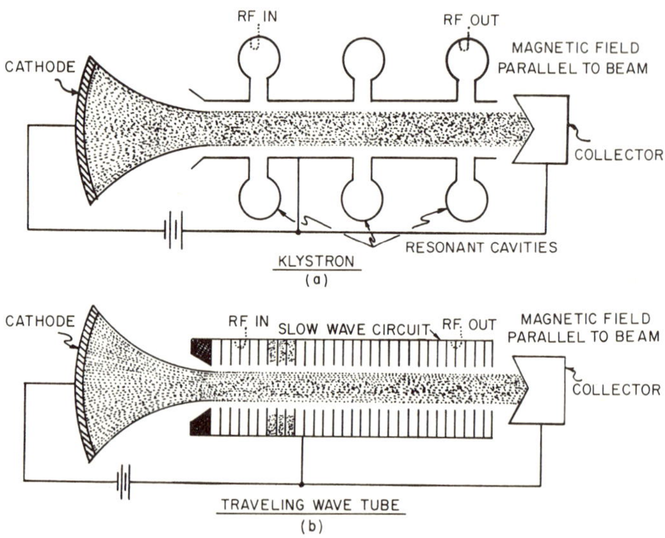

Figure 8. Collinear beam devices. (*a*) Klystron; (*b*) traveling-wave tube.

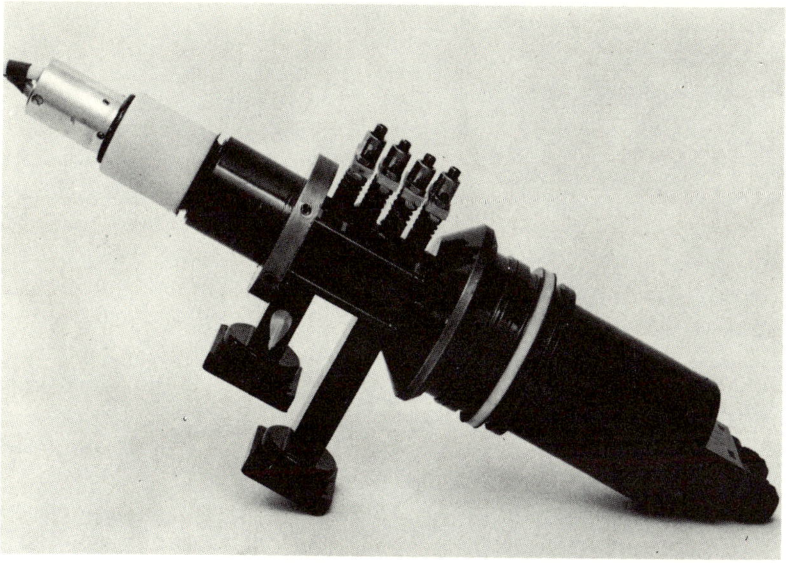

Figure 9. Continuous-wave klystron (VA-849). At a frequency of 8 kMcps, this tube has a gain of 40 db and delivers 25 kw average power, requiring 24,000 volts on the anode. (Courtesy of Varian Associates.)

Traveling-wave tubes (Figure 11) are characterized by good gain, of the order of 30 db, and by excellent broadband characteristics, in excess of 30 per cent of the center frequency.

The second class of high-power tubes is the crossed-field group (Figure 12) in which the electric field for accelerating the free electrons is always perpendicular to a superimposed external magnetic field. These tubes are characterized by the fact that the electrons do not attain the full energy of the applied electric field before interacting. Electrons emitted from the large cylindrical cathode are accelerated radially by the electric field, but are deflected, as they gain speed, by

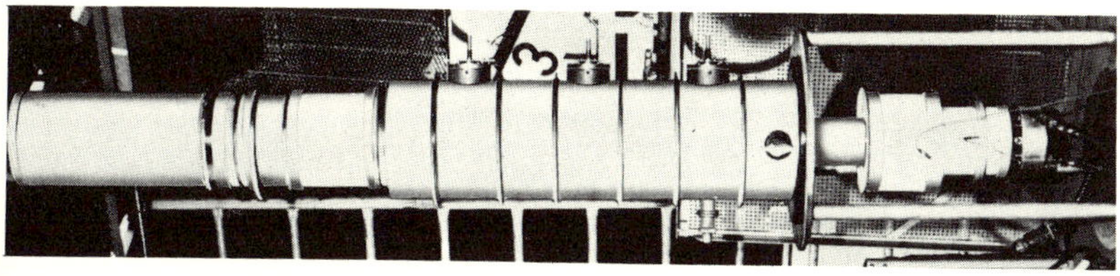

Figure 10. High-power pulsed klystron (X 841). This tube has a narrow-band power gain of 40 db at frequencies of 400 to 450 Mcps, with 100,000 volts on the anode; it delivers peak power of 1.25 megawatts and 75 kw average power. (Courtesy of Eitel-McCullough, Inc.)

the transverse magnetic field. These electrons interact with an incipient field in the anode radio-frequency circuit in a bootstrap operation, whereby the electrons are formed into rotating bunches, much like the spokes of a wheel, while the radio-frequency field on the anode is built up simultaneously.

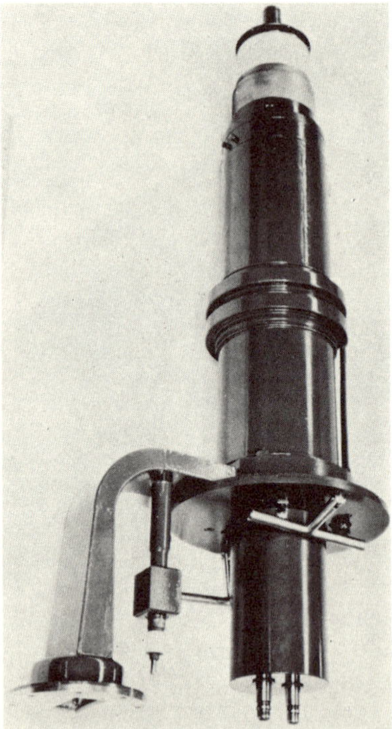

Figure 11. High-power traveling-wave tube (VA-125). In the frequency band 2.68 to 2.98 kMcps, this tube delivers 2.5 megawatts peak power, with a power gain of 37 db, requiring 120,000 volts on the anode. (Courtesy of Varian Associates.)

The magnetron anode circuit is closed on itself; it acts therefore only as an oscillator and not as an amplifier. In contrast, the amplitron has an open anode circuit, and as in the traveling-wave tube, its anode circuit is more of the reiterative type. The radio-frequency circuit of the amplitron is designed so that the electromagnetic wave travels in the opposite direction to the rotation of the electron "spoked wheel"; it is therefore referred to as a backward-wave amplifier.

Crossed-field tubes (Figures 13, 14) generally have greater efficiency (as high as 80 per cent) than collinear field tubes and require lower voltages on the anode. Magnetrons and amplitrons are not limited as to ultimate power capability by thermionic cathode emission. A characteristic of these tubes is the return of the electrons to the cathode, causing the emission of secondary electrons. As the power of the amplitron is increased, a point is reached where secondary emission is adequate for the operation of the tube without externally supplied cathode heating; in fact, the cathode must be cooled rather than heated as in conventional tubes. The power limitation in amplitrons seems to be heat removal from the anode circuit vanes. The use of a large number of very small holes fed with water under high pressure has increased cooling to 15 kw, or approximately 20 hp/cm^2, of anode surface. With an efficiency of 80 per cent, this means that each square centimeter can support an out-

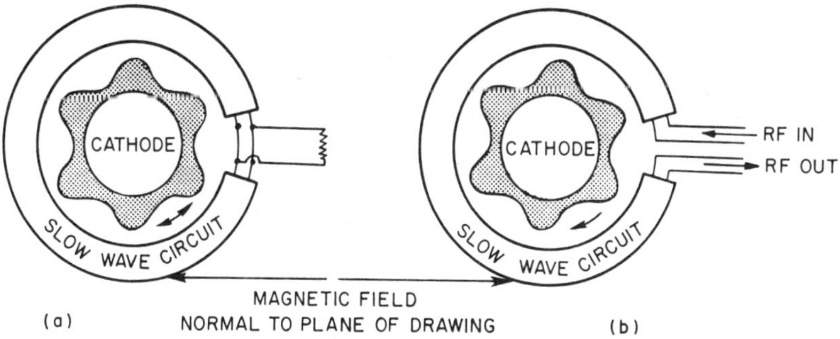

Figure 12. Continuous-cathode crossed-field devices. (*a*) Magnetron; (*b*) amplitron.

put of 80 hp, or a tube of only 20 cm² anode area can deliver about 1,500 hp, as much as a large aircraft engine.

Where do we stand overall in high-power-tube development? The chart (Figure 15) is limited to pulsed radar tubes in the *S* band. Similar charts with different average-power scales can be made for other frequency regions. In general, the power varies directly as the square of the wavelength; hence at *X* band, one-tenth of the average power shown is available, and at *L* band, ten times the average power shown.

Figure 13. High-power *X*-band magnetron (SFD 303). At a frequency of 10 kMcps, this tube delivers peak power of 1 megawatt and average power of 1 kw. (Courtesy of Saloom, Feinstein, Drexyl, Inc.)

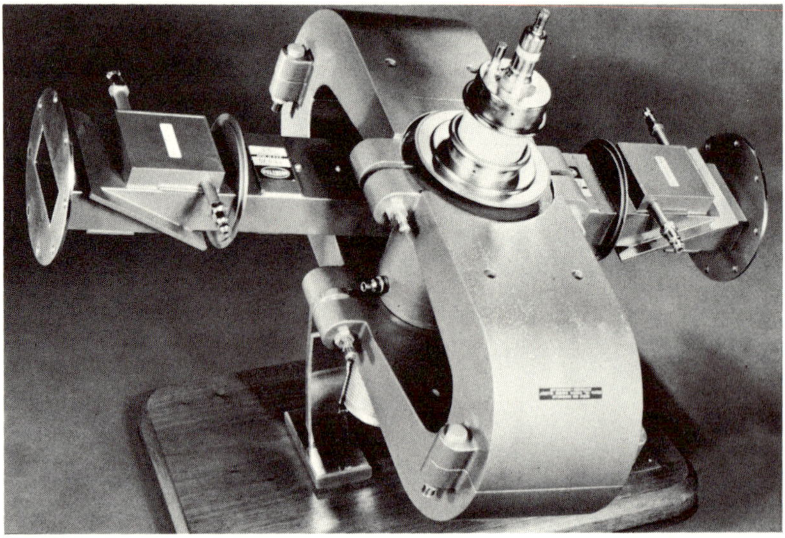

Figure 14. High-power S-band amplitron (QK 622). As an amplifier, this tube has a gain of 10 db and a bandwidth greater than 6 per cent at a frequency of 3 kMcps; it delivers peak power of 3 megawatts, average power of 15 kw, with an efficiency above 70 per cent. (Courtesy of Raytheon Co.)

Today, average power at 12 kw is readily available at 3 kMcps. There is no reason why the same principles and technology should not bring these levels up to 300 kw or more, that is, 500 hp at S band, generated in a volume smaller than one-tenth of a cubic foot. In fact, as we look to the future, it appears as though the limits of the application of high-power microwave energy will not be due to the transmitting tube, but rather to the cost of the ancillary equipment, so that there seems to be no technical bar to providing microwave power in the neighborhood of 10 megawatts.

As we go toward millimeter waves, the difficulty increases, but there is hope that they will be solved through the use of high magnetic fields produced by superconductors and through the combination of large numbers of amplifying elements into parallel assemblies. Beyond the millimeter region we enter infrared and visible light, and here the maser and laser come into their own.

The performance of a radar also depends on the sensitivity of the receiver. But whereas there is no theoretical limit on high power, there are limits to useful receiver sensitivity. In fact, improvement in sensitivity beyond a certain point is often undesirable, because of natural or man-made radio noise. Examples of each of these are well known.

Lightning is always accompanied by radiation of radio noise of all frequencies, and the same ionosphere which makes possible long-range low-frequency voice communication also helps the transmission of these same frequencies of radio noise around the world. Thus the large number of lightning storms raging all over the world add to produce a noise level which depends principally on the condition of the ionosphere and which is higher at the lower frequencies. This level of noise is so high that in the 1930s there existed no problem in making low-frequency communication receivers adequately sensitive. Increasing sensitivity increased the noise level along with the signal level.

Over and above the natural noise level, man makes his own radio noise; an example is the radiation from automobile ignition systems. One of the most important functions of the Federal Communications Commission is to define standards of extraneously radiated radio noise and to enforce these standards on equipment manufacturers and users.

Noise can be described quantitatively in terms of absolute temperature. If an oven is raised to a uniform temperature, the space within the oven will also be at that temperature, and any substance placed within will in time come into temperature equilibrium with the oven. Under these conditions, the entire oven is filled with what is called black-body radiation. The energy density at each frequency interval

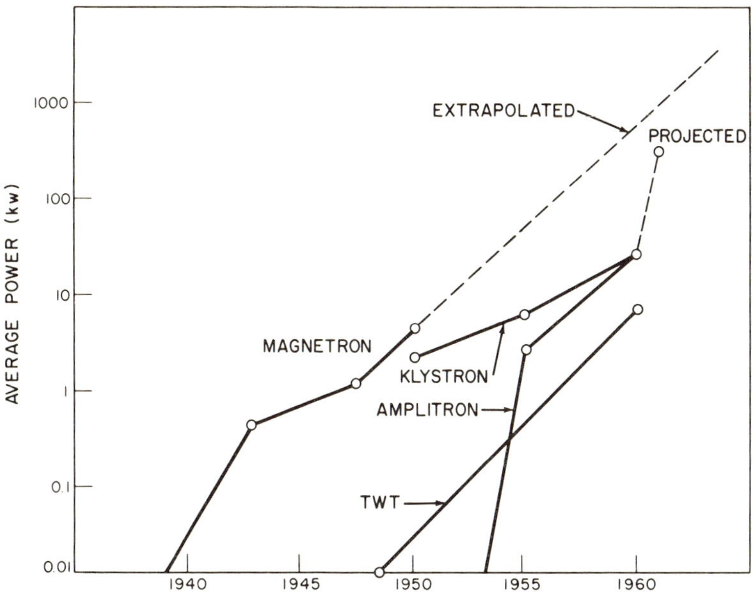

Figure 15. Availability of power in S-band (3-kMcps region) pulsed tubes. (Prepared by William Brown.)

of this radiation was first correctly stated by Max Planck and is uniquely defined by the temperature. Conversely, if one measures the energy density within any frequency interval, one can determine the temperature of the oven. It must be reiterated that the thermodynamic definition of temperature requires a condition of temperature equilibrium. Nevertheless, the idea of using temperature as a measure of electromagnetic energy is a useful one. For example, the radio noise at one megacycle on the earth corresponds to a temperature of 100,000,000°K. This is not to say that the earth is very hot, but there are radiation sources which apparently radiate selectively as much energy at this frequency as would be found in an oven at 100,000,000°K. Of course, if the energy density at all other frequencies were related to the energy density at 1 Mcps by Planck's law, then indeed, we all should be in an oven at 100,000,000°K.

The noise temperature of the atmosphere is very high at low frequencies, well over 100,000,000°K (Figure 16). It is not until about 25 Mcps that atmospheric noise is equal to the radio waves radiated by material at room temperature, 290°K. The sky also appears hot, and not until approximately 200 Mcps does the sky look to a radio receiver like material at room temperature. Actually, outer space near the earth is at a temperature of about 4°K, but this sky temperature is not reached until about 100,000 Mcps.

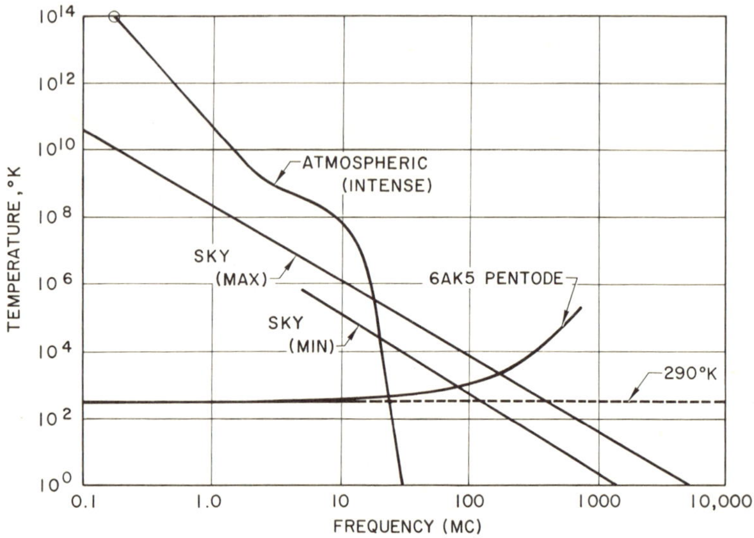

Figure 16. Atmospheric and galactic noise relative to that of a conventional amplifier (6AK5 pentode).

If we compare these noise levels with the noise inherent in a receiver using a conventional 6AK5 pentode tube, the conclusions are quite obvious. In low-frequency communication, the 6AK5 does not contribute to noise and there is no useful point to be served by improved design. On the other hand, at radar frequencies of 500 Mcps and above, the background noise drops to below room temperature and the 6AK5 seriously limits performance.

Historically, the development of low-noise receivers followed the availability of ever-higher-frequency transmitter tubes. Again, it was the amateur exploitation of high-frequency radio in the 1930s that supplied the impetus for the so-called acorn tubes, and later the Lighthouse triodes, so named because of their shape. But the equivalent noise temperatures of these tubes at microwave frequencies were extremely high. Strangely enough, the old cat's-whisker crystal of the Quaker Oats box radio of the 1920s was resurrected and saved the day. The crystal converter is still the workhorse of radar receivers. With improvements in IF tubes and in coupling, it is now possible to realize at 10,000 Mcps an overall receiver, using crystal converters, with an equivalent temperature of only 1000°K.

The last ten years have been marked by a large number of breakthroughs in low-noise amplifiers. By sheer effort, the triode, the 416B, has been made useful up to 4,000 Mcps. The traveling-wave-tube principle has also responded to careful analysis and design. By controlling the electron beam at low velocity, the noise inherent in the statistical spread of the beam has been reduced. Combined with clever circuitry, this has brought the traveling-wave tube to a high level of performance, comparable to that of the best crystal mixers and microwave triodes (Figure 17).

Two other devices, the semiconductor parametric amplifier and the maser, have made possible revolutionary improvements in low-noise receivers. The parametric amplifier uses germanium or silicon diode as a variable capacitance. Parametric amplifiers are not new in concept; they have been known to electrical engineers, in one form or other, for over a half century. In the simplest form, any tuned circuit can be furnished energy by varying at twice the nominal resonant frequency the parameter of one of the resonating elements. This is essentially the action of a child as he shifts his weight to the back on the forward half of his swing. Once the swing is set into motion, the child amplifies the resonant frequency by pumping. In the electrical equivalent, the inductance or capacitance, or both, is changed. In the microwave amplifier, the trick is to find a mechanism which permits variation of inductance or capacitance, or both, at twice the desired frequency.

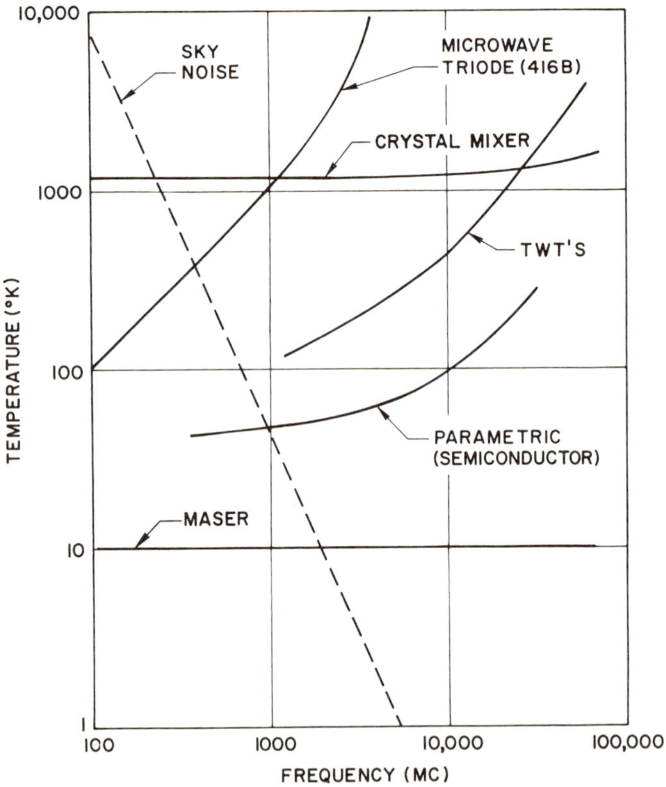

Figure 17. Low-noise receivers. State of the art, 1961.

Obviously, this cannot be done mechanically at microwave frequencies. Fortunately, the capacitance of a semiconductor diode is a function of applied voltage. If, therefore, a sufficiently high voltage at twice the desired amplification frequency is applied, weak signals can be amplified. The semiconductor parametric amplifier produces quite good noise figures with relatively simple circuitry and without refrigeration. Equivalent temperatures as low as 70°K have been demonstrated. With refrigeration of the semiconductor to liquid-air temperatures, noise figures of 50°K have been produced.

The maser is the most glamorous of the new devices. In principle it is capable of almost noiseless performance at microwave frequencies. The fundamental limitation is noise due to spontaneous emissions of photons, independent of the emissions stimulated by the incoming signal. These spontaneous emissions are not coherent with the signal being amplified. In degrees, the basic maser noise temperature is equal to $0.05f$, where f is kilomegacycles per second. At 10,000 Mcps, the noise temperature is only 0.5°K, but associated circuitry inevitably raises the effective temperature to about 100°K in current experimental radars and about 8°K in radio-astronomy receivers.

Whereas in 1940 there were no satisfactory low-noise amplifiers in the 3-cm, or 10,000-Mcps, region, the designer now has a wide choice (Figure 18). In general, it can be said that radar performance in the future will not be substantially improved by better receiver design. Industrialization, the ever-increasing use of electronic aids and communication, and, in military applications, potential enemy jamming will all tend to reduce the usefulness of higher sensitivity. Improvements in receivers will come more from simplification, particularly with the capability of rapid frequency tuning, broadbanding, and reduction of ancillary gadgetry.

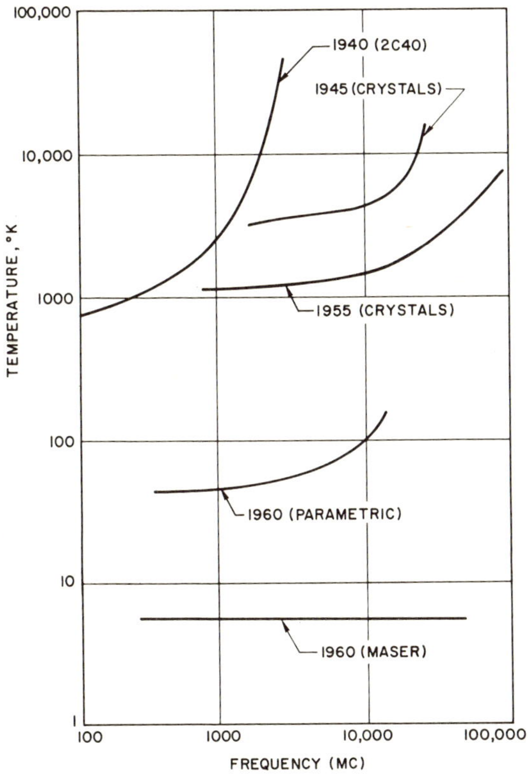

Figure 18. Progress in low-noise receivers, 1940–1960.

We now turn to another set of problems which face the designer of any radar equipment: that of target discrimination or contrast. The human eye can see a specific object in contrast to all the other background objects through the utilization of various factors: color, shape, motion, angular position, and depth perception. The same general processes are available to the radar designer, but when considered in detail, the applicability to specific problems is not always obvious. In visual detection, we are accustomed to frequencies which are generally 25,000 times as high as microwave frequencies. As a result, the angular resolution of the eye, in spite of its relatively small aperture, is manyfold higher than that of a radar system. Modern meter-wavelength radars with antennas hundreds of feet in diameter have beams approximately one degree wide. Obviously, to obtain higher resolution in reasonably sized antennas, one must go to higher frequencies. Although our technology now can support transmitters and quiet receivers in the millimeter-wavelength region, the utility of these high frequencies is limited by atmospheric absorption and by Rayleigh scattering in rain similar to the scattering of visible light by fog or

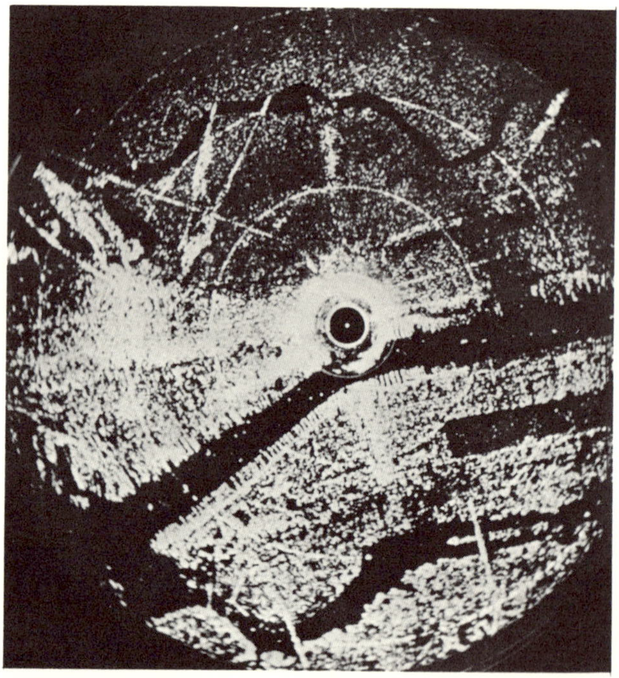

Figure 19. High-resolution map of Manhattan Island, New York City, made with airborne M.I.T.–Radiation Laboratory World War II experimental equipment.

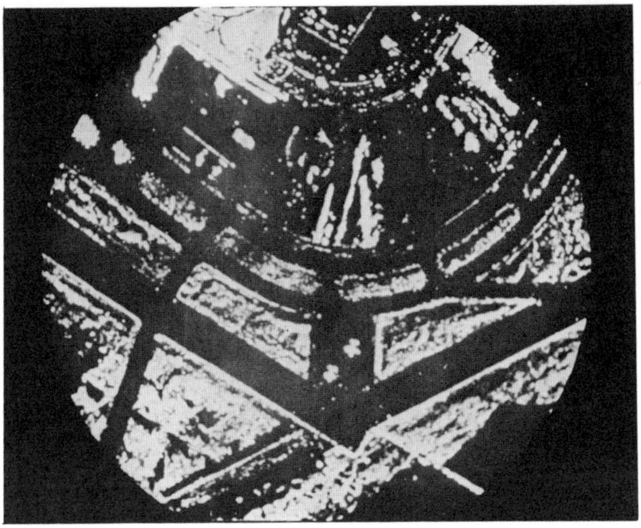

Figure 20. Radarscope. Airport Traffic Control radar (ASDE), Idlewild Airport, New York. (Courtesy of Airborne Instruments Laboratory.)

clouds. The practical limit on the earth, but of course not in space, is about 1 cm wavelength, or 30 kMcps. At these frequencies, an antenna only a few feet in diameter results in a beam of $\frac{1}{4}$ degree.

Complementary to angular resolution is range resolution. Since a pulse one-millionth of a second long corresponds to about 500 ft, geometry, or shape recognition, is clearly feasible (Figure 19). The potential of this sort of equipment in navigation and bombing is obvious. Ground equipment at *K*-band frequency, or 1 cm wavelength, produces a beam $\frac{1}{4}$ degree wide and, with pulses only $\frac{1}{50}$ of a millionth of a second long, provides a resolution of 10 ft. Used for airport traffic control (Figure 20), two airplanes in a holding position can be clearly distinguished, and an experienced operator can easily identify different plane types.

Target contrast by color has not proved as useful to radar as it has to the eye. The reflection of radio waves is principally a macroscopic process, whereas the reflection of light results primarily from atomic or molecular interaction. Hence most objects do not selectively reflect radio or radar frequencies. Perhaps the greatest improvement in terms of target contrast has been through discrimination of objects which are moving against a background of objects that are at rest, the situation that applies to the detection and tracking of fast-flying airplanes in the presence of ground reflections.

The simplest method of velocity discrimination depends upon interference of the reflection of the moving target with the other signals. As a result of phase differences due to changes in position of the target, its reflected signal will alternately be added to, or subtracted from, the other signals, thus causing a fluctuation in the total signal received. This was essentially the effect first noticed by Naval Research Laboratory workers in September 1922. It is the cause of the fluctuation of television pictures as an airplane flies by. It was used in World War II to detect troops moving at night, and is the basis of radar burglar alarm systems.

The steady motion of a target causes a shift in the frequency of a reflected radio signal; this is the well-known Doppler effect, which also accounts for the change in pitch of a whistle or horn as a vehicle passes by. Prior to World War II, much effort was expended on continuous-wave systems, using this Doppler effect to distinguish moving airplanes against the large, overpowering reflections from the ground.

But the then clear advantages of pulsed radar, with resultant simple range discrimination, threw the massive effort into pulsed technology,

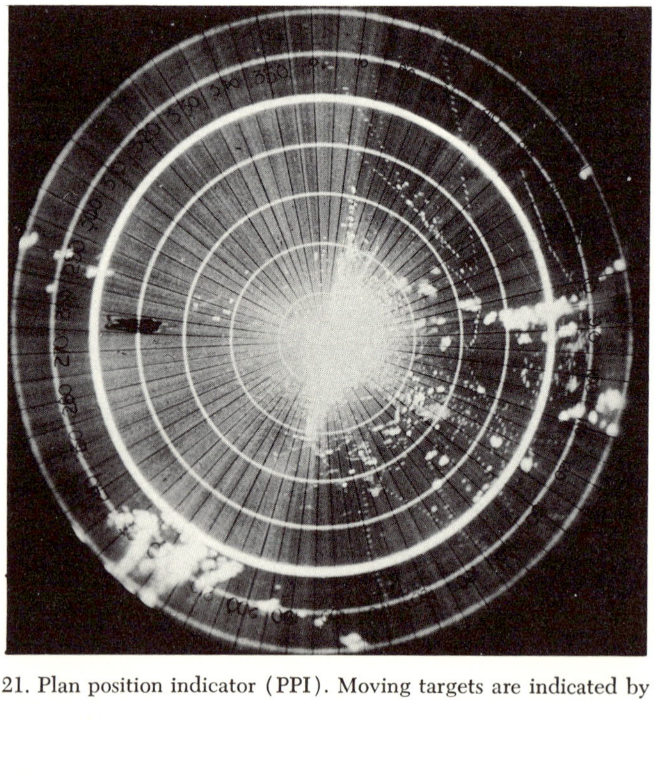

Figure 21. Plan position indicator (PPI). Moving targets are indicated by tracks.

so that with the exception of radio altimeters, essentially all radar till 1950 was of the pulsed type. The problem of velocity discrimination then reduced itself to superimposing such capabilities on the pulsed equipment. In the simplest embodiment, this is done, not by using the Doppler effect, but by repeated looks at a plan position scope picture as shown in Figure 21. The airplanes are clearly detectable from fixed target by the tracks. This can be done photographically, by a storage tube, or by a digital computer of adequate size.

Unfortunately, reflections from large ground objects and from wet clouds can mask the airplane signals completely.

Ground clutter can be reduced by MTI (moving target indicator) or by pulse-Doppler techniques. These techniques involve examining all signals and comparing them with a stable oscillator which is locked in (i.e., is coherent) with the transmitted signal. A fixed target then reflects signals whose phase relative to the stable oscillator is invariant. The relative phase of the signal from the moving target changes from pulse to pulse, and by comparing these phases with the stable oscillator, the successive returned signals will appear to fluctuate.

The moving-target-indicator technique cannot cope with all types of clutter. For example, wet clouds reflect radar signals, particularly at the higher frequencies. Furthermore, clouds often move at speeds in excess of 60 mph, a speed greater than the radial component of many fast-flying airplanes. Hence MTI in itself is often insufficient in suppressing cloud clutter. Advantage can be taken, however, of the fact that water droplets are small spherical scatterers, rather than specular reflectors. Thus, with appropriately designed antennas using circular polarization, the reflections from droplets can be reduced to 1 per cent, with only a small degradation in wanted aircraft signal strength.

Since the signal strength of a radar varies as the fourth power of range, it is obvious that the sensitivity at, say, 4 miles is 10,000 times too high as compared with 40 miles. STC (sensitivity time control) overcomes this overloading at near ranges by varying the gain, in this case by a factor of 10,000, during the thirty-millionths of a second corresponding to the delay between the arrival of signals from 4 and 40 miles. The overall effect of circular polarization, MTI, and STC is quite dramatic (Figures 22, 23).

Finally, consideration may be given to "angels," so named by the British in World War II during the radar watch over the Channel. Angels are any signals of unknown origin; many such signals have now been identified as reflections from birds. Since even a small airplane is

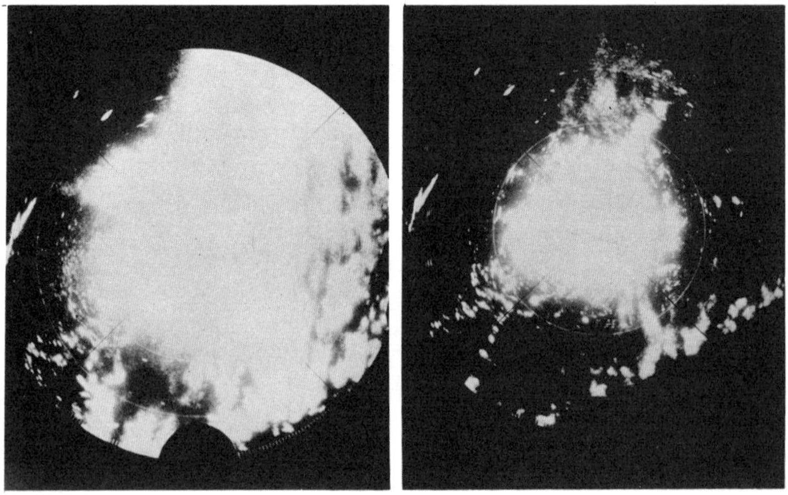

(a) (b)

Figure 22. Plan position indicator of ARSR-1 in-flight control radar. The picture, taken during a storm, consists of three superimposed scans, 10 min apart; thus an airplane should appear as three separated dots. (*a*) With linear polarization: the PPI face is almost entirely covered with clutter; (*b*) with circular polarization: cloud clutter is greatly reduced, but ground clutter from mountains remains.

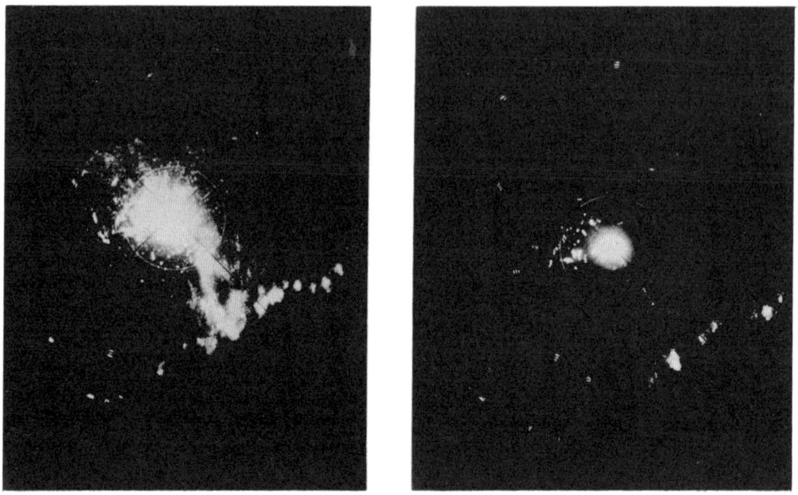

(a) (b)

Figure 23. Plan position indicator (same as Figure 22). (*a*) With circular polarization and MTI: ground clutter is reduced substantially; (*b*) with circular polarization, MTI, and STC: airplanes are now clearly detectable. (Courtesy of Raytheon Co.)

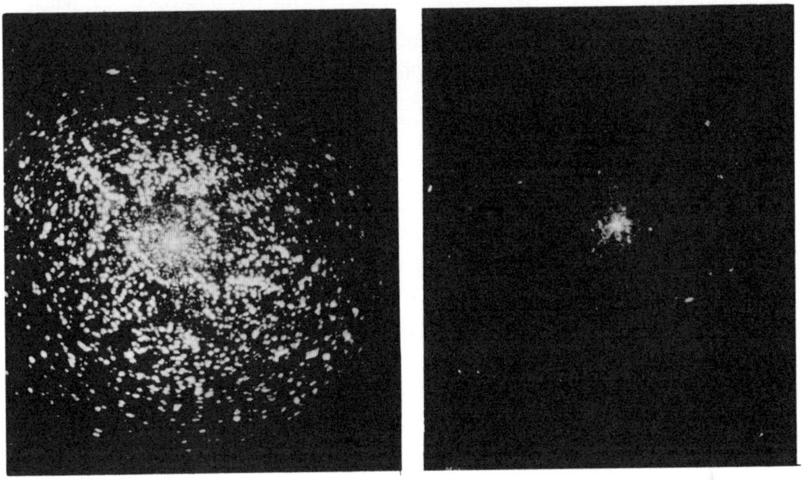

(*a*) (*b*)

Figure 24. Plan position indicator of ARSR-1. (*a*) Clutter due to reflections from birds (crows, hawks, and buzzards); (*b*) bird clutter eliminated by STC. (Courtesy of Raytheon Co.)

some 200 times as large a reflector as a bird, the STC can be set to eliminate bird reflections at all ranges, while not eliminating those from airplanes (Figure 24).

These techniques were substantially understood in 1950. But there are limits to the amount of ground-clutter rejection possible. Part of the limitation comes from the limits of stability of the transmitter and local oscillator; part comes from the fact that most radars must scan, so that even fixed targets change in signal strength as they move through the beam; part comes from the fact that the pulses are repeated at some rate depending generally on the maximum range of the radar. This finite pulse rate causes blind spots in the velocity discrimination and forces the designer to higher and higher repetition rates, as the closing speed of the target is increased. This problem is particularly aggravated in airborne radars.

An interceptor aircraft or missile, flying at Mach 3.0 toward an approaching enemy plane also flying at Mach 3.0, would require an *X*-band radar with a pulse rate around 300,000 per second, or an interval between pulses of only 3 microseconds, corresponding to about $\frac{1}{3}$ mile. With such a high repetition rate, range measurement becomes ambiguous. Disregarding for a moment the technical problems associated with high repetition rates, it is clear that at these high rates the duty cycle will approach 50 per cent and the peak power will be only twice the average. We have nearly closed the loop back to continuous waves. Nevertheless, such equipment has been designed, and the broad spectrum of pulse-Doppler equipment developed during the past ten years marks a tremendous advance in closing the low-altitude gap of air defense.

But these last ten years have also seen a tremendous growth in continuous-wave radars. Continuous-wave radars do not lend themselves, like pulsed radars, to simple-range discrimination and multiple-range indication, such as provided by pulsed radars with the plan position indicator. On the other hand, the Doppler effect makes CW radar attractive where moving targets are to be picked out and the range of the one or few of interest is then to be determined. Thus the CW approach is particularly attractive in the problem posed by airplanes flying close to the ground. Practically no work was done on CW radar during the war or for the five years after; the low-altitude enemy airplane attack remained the one major hole in our air defense.

A moral was drawn earlier: when facing a new situation, always go back to first principles. The development of CW radar required the development of new transmitting tubes, new types of receivers, new kinds of waveguides, and basically different design criteria. Early models of CW radar had very poor performance, principally because of the large amount of spillover signal from the transmitter to the receiver and also because available transmitting tubes had both amplitude- and frequency-modulated noise, which mixed with the carrier frequency to drown out the weak reflected Doppler signals. Both of these limitations have been removed during the past ten years. This new technology is now incorporated in four of the nation's major antiaircraft missile programs.

Information theory has important applications in radar. The principal functions of radar are, first, the sensing of information, and then the use of that information. Each of these areas is major in scope; I shall limit my comments to two examples of signal enhancement.

Signal integration is most simply described by considering a pulsed-signal return (Figure 25). A signal may appear to be lost in noise. If, however, repeated signals are added over successive returns, then the true signal will add linearly, whereas the noise, being random from pulse to pulse, will add as the square root. This simple application of information theory was in use in automatic-range-tracking sets during World War II.

Recently, more elaborate application has been made of this technique at Lincoln Laboratory to detect the return radar echo from the

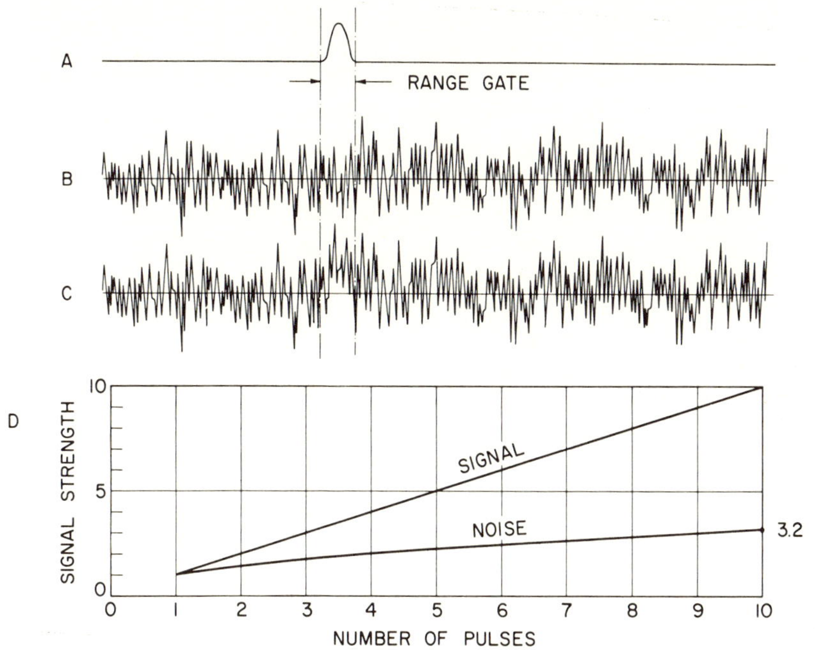

Figure 25. Signal integration. (*A*) Ideal video signal; (*B*) noise; (*C*) signal superimposed on noise; (D) summation of signal and noise over successive pulses.

planet Venus, some 28,000,000 miles away. The radar was the huge Millstone Hill tracking radar (Figure 26). This antenna is 84 ft in diameter and is steerable, so that pointing of the narrow beam toward Venus is readily accomplished. With its high peak power of 2.5 megawatts, long pulse length of 2 msec, and ultramodern maser receiver with overall noise temperature of only 240°K, the radar was still not adequate to obtain detectable reflections in a single pulse. In fact, an increase of some 100 times in power would be required before a single-pulse echo could have been detected even by cross-correlation of that pulse. This approach was clearly impossible.

Instead, the Lincoln Laboratory scientists effectively stretched the outgoing signal by transmitting a train of short pulses spaced at varied, but controlled, intervals of time. The transmitted signal had the appearance of a long periodic sequence of these short pulses, at a regular rate of 30 pulses/sec, except that half of the pulses were deleted in an irregular but controlled manner, to create a special time-sequence pattern which could be used by the computer to pick out the weak returns. The pulse train was transmitted for slightly under 5 min.

Figure 26. Millstone Hill tracking radar, Westford, Mass. (Courtesy of M.I.T.-Lincoln Laboratory.)

Then, as the transit time to Venus and back, even at the velocity of light, is about 5 min, the receiver was activated in time to receive and record the complete train of echoes. Several modes of signal processing were used, including coherent integration for approximately 1 sec, followed by video integration for 5 min, and video integration over many successive 10-min transmit-receive cycles (Figure 27). This required correction for the receding velocity of Venus from the earth, approximately 15,000 mph.

These experiments represented two major areas of accomplishments: the successful application of off-time information processing in radar echoes and the astronomical implications. A new measurement has been made by radar of the size of the solar system—a measurement which is two orders of magnitude more accurate than any previous optical determination.

4

Today we stand on the threshold of space. We took our first step with the intercontinental ballistic missile, and we took a giant leap when the first artificial satellite went into orbit, on the 4th of October 1957. No one is wise enough to predict the future clearly. We already know that space is not a friendly environment for man; extended periods of weightlessness and strong particle-radiation fields confront the intrepid adventurer. But regardless of whether man himself goes into space, certainly man will exploit space to his needs, be they peaceful or warlike. Communication satellites, weather satellites, and other satellites will become common and affect our ways of life. In all these, radar and its immediate cousin radiotelemetry and tracking will play important roles.

History has a strange way of repeating itself; the stage setting is somewhat different and the actors are all new, but the themes are not. Twenty years ago, in response to the Nazi military airplane threat, a generation of radars was developed. Today, in response to the communist intercontinental ballistic missile, there has been developed the Nike-Zeus system, with radar and associated data-processing equipment to provide accurate tracks of moving targets in its full hemisphere of coverage.

In 1940, an antenna the size of a billboard taxed the ingenuity of a designer. Today its counterpart against intercontinental ballistic missiles is the BMEWS (Figure 28). Each BMEWS antenna is bigger

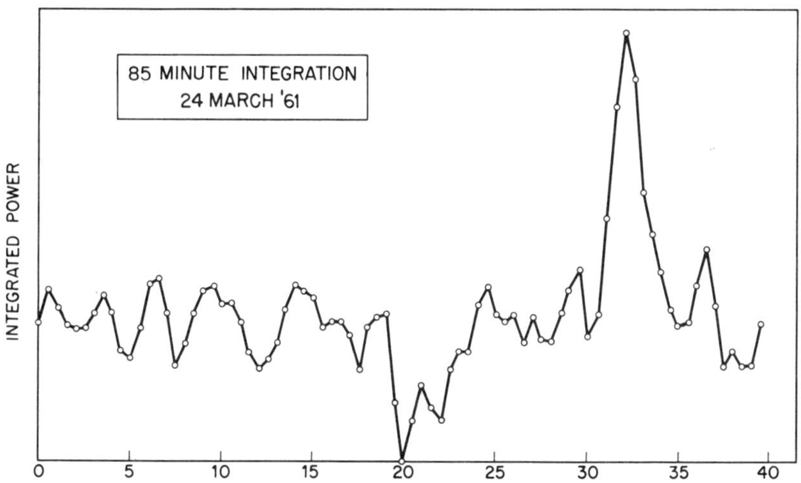

Figure 27. Received signal intensity vs. range over the interpulse interval, for radar echoes from Venus, 24 March 1961. A 2-msec pulse width was used.

Figure 28. Ballistic Missile Early Warning System (BMEWS) radar, Alaska. (Courtesy of Radio Corporation of America.)

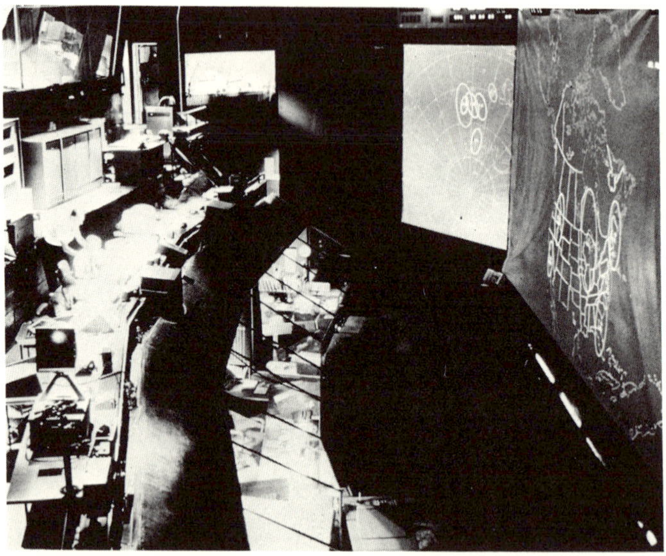

Figure 29. BMEWS radar display at NORAD (North American Air Defense Command), Combat Operations Center, Colorado Springs, Colorado. (Courtesy of Radio Corporation of America.)

(*a*)

(*b*)

Figure 30. (*a*) Hypothetical satellite rendezvous system. The satellite has just been separated from its booster and is being ejected into orbit. The shrouds, which protected it from aerodynamic heat in the atmosphere, are separating. (*b*) The closing phase of the rendezvous is accomplished by a tracking radar in the satellite.

than a football field. Arctic winds and snow and the logistic problems of isolated posts no longer daunt the intrepid engineers and soldiers. The improvised combat control center of World War II has been replaced by the modern operations room (Figure 29). In this modern version, the plots cover continents but time has shrunk to minutes.

Beyond the intercontinental missiles lies limitless space with satellites dispatched accurately to rendezvous at predetermined points (Figure 30*a* and *b*) and, in the future, space ships traveling beyond the earth's gravitational field. In space, the radar designer is unencumbered by air or gravity. Without an atmosphere to absorb radiation, any frequency may be used. Thus millimeter waves, radars, and light radars using optical masers, all of which have as yet little or no practical applications on the earth, will be developed.

In retrospect, these last ten years since the founding of Lincoln Laboratory have been challenging, and much has been accomplished. We live in a world of explosive technology. The next ten years will be critical indeed, but we can confidently expect that the strength of our nation will be enhanced by our qualitative leadership in radar and in electronics.

REFERENCES

1. A. P. Rowe, "One Story of Radar," p. 9, Cambridge, London, 1948.
2. Henry Guerlac, The Radio Background of Radar, *J. Franklin Inst.*, vol. 250, no. 4, p. 286, October, 1950.
3. Ref. 2, p. 295.
4. Ref. 1, p. 35.

Stanislaw M. Ulam, Research Adviser at the Los Alamos Scientific Laboratory of the University of California, Los Alamos, New Mexico, is a mathematician noted for initiating the Monte Carlo method for the statistical treatment of mathematical and physical problems too complex for conventional analysis.

CHAPTER 5

ELECTRONIC COMPUTERS AND SCIENTIFIC RESEARCH

1

The advent of large computers has created apprehension among some mathematicians and scientists. It is feared that individual scientific efforts will be pushed into the background or replaced by less imaginative, purely mechanical habits of research. Such fears seem to me quite groundless. The computer can be looked upon as an extension of a very old and simple invention—a stylus or a pencil. Even the most abstract thinkers agree that the simple act of writing down a few symbols on a piece of paper helps enormously in the thinking process. Concentrating on what is visually in front of one, rather than relying exclusively on the patterns in the memory, seems to be very useful. In this respect alone, and it is not a trivial one, the new electronic machines provide a marvelous extension of the use of symbols in science, increasing our effective memory. Furthermore, computer display devices enable one to visualize patterns, be they ever so abstract, in a way which extends the range of the *Gedanken Experimente* that have given rise to so many new ideas in mathematics and pure physics. In presenting some aspects of the new art and outlining the possibilities of novel experiments, my point of view is that of a customer, a user of this new research tool, the electronic computer.

It is significant that some of the earliest ideas and inventions in the field of automatic calculation were due to people with an extremely abstract turn of mind. Pascal produced a calculator. Continuing through Babbage in the nineteenth century and ending with the mechanical relay devices of Vannevar Bush at M.I.T., one sees enormous progress in the techniques attempting to realize a program of Leibnitz. But it is essentially at the end of the Second World War, when new developments in electronics made possible much faster and more compact machines, that the new era may be said to have started. Relatively large memories and the use of a general code increased greatly the variety of problems which could be handled by the machine. Von Neumann's contributions were especially important in this respect. While the systems of wired connections remained fixed, a diversity of codes for different problems on the same physical background enabled one to treat a wide range of mathematical schemas.

The technology of the machines themselves is still in rapid development. During the last decade, the speed of the elementary operations that the machine performs (simple logical orders, additions and multiplications of real numbers) has increased by a factor of about 100. Modern computers can multiply two numbers, each of 40 binary digits or so, in 1 microsecond or less. The size of fast-access memories has also increased by a very large factor. The readily accessible memory is of the order of 100,000 words; there is no essential limit to the size of a memory with slow access.

Some of these characteristics will no doubt be further improved during the next few years. Rapid progress in miniaturization techniques hastens the trend toward smaller, more compact machines. Parallel to this development is an improvement in the variety of new logical and mathematical orders concerning Boolean and other operations. More recently, auxiliary and ancillary equipment is becoming available which permits a visible display of the results obtained by the computers. At the same time, work is proceeding on easier methods of insertions or changing the data going into the calculation or, indeed, changing the course of the calculation itself, on the basis of the results produced and displayed by the machine. This will bring about a more intimate cooperation (if one may properly use the expression) between scientists and programmers, on the one hand, and the machine, on the other.

2

It is known that Gauss greatly favored experimentation. When he was asked how he managed to divine some of the remarkable regularities of numbers, he answered, "*Durch planmässiges tattonieren,*" through systematic trying. This may well be considered a definition of the pragmatic method in mathematical research. Its modern application, using electronic computers, has already produced a vast body of literature, from which one chooses for purposes of illustration not necessarily the most important examples, but those which are familiar and, perhaps, more personal.

As all mathematicians know, great mysteries remain in number theory, despite the logical simplicity of the subject; there are very simple questions, explainable to a ten-year-old, which are still unanswered. Many of these concern properties of primes. It has proved fruitful to examine the *appearance* of primes as written in the binary system; this is a quick and simple process, using computers. One tries to obtain by inspection some general rule regarding the distribution of 0's and 1's in the binary expression of a prime integer. The so-called Fermat primes are of the form $2^{2^n}+1$ (Figure 1). By means of a computer, one easily develops a long list of such integers. It is seen that Fermat primes can be described as prime integers whose expression in binary notation contains two 1's, the rest of the digits being 0's. It is conjectured that there are infinitely many Fermat primes.

One could ask a seemingly simpler question: Are there infinitely many primes in whose binary expression there is any fixed number, k, of 1's? Even this weaker conjecture seems very hard to prove. At Los Alamos, we have used a computer in statistical studies of the combinatorial properties of the distributions of the 0's and 1's in primes. One day, Dr. Mark Wells, who was working with me, said, "Of course, one cannot expect the primes to have asymptotically the same number of 1's and 0's in their development, since the numbers divisible by 3 have an even number of 1's." Taking this remark as representing some very easily provable elementary fact, I returned to my office. But then I found, after considerable effort, that I could not establish it. In fact, the statement is not even true (Figure 2). The first integer

n	Decimal Notation	Binary Notation
0	3	11
1	5	101
2	17	10001
3	257	100000001

Figure 1. Fermat primes.

Decimal Notation	Binary Notation
3	11
6	110
9	1001
12	1100
15	1111
18	10010
21	10101

Figure 2. Prime integers divisible by 3.

divisible by 3 which has an odd number of 1's in its binary expression is 21.

Nevertheless, a great majority of the integers divisible by 3 seem to have an even number of 1's. Beginning with this observation, Wells managed to prove a general theorem: among all the integers divisible by 3, from 1 to 2^n, those which have an even number of 1's always predominate, and the difference between their number and the number of those with an odd number of 1's can be computed exactly and is of the order of $3^{(n-1)/2}$. It is even more interesting that he developed corresponding proofs for integers divisible by 5, 7, and some other numbers, although these theorems become harder and harder to prove. A paper describing this work will appear in a mathematical journal. This provides a very nice, if modest, illustration of the claim that an examination of special cases, yielded quickly and in abundance by a computing machine, may stimulate a mathematician to conjecture, and then perhaps to prove a general truth.

Professor S. Chowla, a famous number theorist now at the University of Colorado, mentioned to me recently his interest in the equation

$$\binom{l}{3}+\binom{m}{3}=\binom{n}{3}$$

where l, m, and n are integers. For example,

$$\binom{5}{3}+\binom{5}{3}=\binom{6}{3}$$

That was the only case of a solution in integers which he knew, and he thought that this was the only one possible. As a result of the brief discussion between us and a young student at the university, we decided to investigate the problem on a computing machine. It turned out that, even for moderate values of n ($n < 100$), there were many other solutions. Inspection of the special nature of these gave Chowla an approach to the proof that there are infinitely many such triplets. He also proved there is only one solution, the one stated above, when l and m are equal.

It seems to be true that mathematicians are not happy unless they can prove that something exists but is hard to find, or else that there are infinitely many members in a given set. But it is significant, as in this case, that the infinity of solutions was suggested by the finite process carried out on the machine. One may hope that not only isolated curiosities, but hints about general facts, will be obtained through empirical work followed by thought. D. H. Lehmer in California has shown the great value of computing machines in number theory research.

Other illustrations of the pragmatic method can be found in certain topics of what can be called elementary algebra. It is perhaps not generally realized how little is known to mathematicians about transformations which are not linear. In a space of, say, three real variables, x, y, and z, one can consider a transformation

$$x' = f_1(x, y, z)$$
$$y' = f_2(x, y, z)$$
$$z' = f_3(x, y, z)$$

where f_1, f_2, and f_3 are not linear functions, but quadratic functions, of these variables. Almost everything is known about the linear case, but if quadratic or higher-order transformations are involved, some of the simplest questions concerning the properties of such transformations and the behavior of their iterations remain unanswered.

It is more than mere curiosity to inquire about these; there are indeed practical reasons for studying nonlinear transformations and equations. The laws of nature do not require that fundamental relationships involve only linear operators. Despite the success of the well-established (linear) quantum theory, there are now indications that in certain basic aspects it is essentially nonlinear. Of course, in classical theories, such as hydrodynamics, the equations are nonlinear from the beginning. Hence it is useful to give a brief account of a systematic investigation, performed with the aid of computers, of some elementary mathematical problems concerning quadratic transformations in the space of two, three, and four dimensions.[1]

Let us imagine a population of n particles of, say, three different types, which may be thought of as different colors. The number n is very large. We assume that these objects collide or mate in pairs, at random, and that each pair will produce a pair of offspring for the next generation, the color of the offspring being uniquely determined by the colors of the two parents, according to a fixed rule; i.e., the characteristics of the offspring are a function of the characteristics of the two parents. Many such rules are mathematically possible. We shall not restrict ourselves to those that might make biological sense, but consider the problem in its full generality. Suppose x, y, and z denote the proportions of the particles of each type in the original population. Then $x + y + z = 1$. Given a rule of inheritance of the colors for the next generation, the question is: How will the fractions change in the course of time, that is to say, with the passage of generations?

The new compositions of the population could be, for example,

$$x' = y^2 + z^2$$
$$y' = 2xy + 2xz$$
$$z' = x^2 + 2yz$$

This corresponds to the following scheme: Two y particles produce a particle of the x type. So do two z particles. The collision of an x particle and a y particle, or an x particle and a z particle, produces a y particle. The mating of two x particles, or a y particle and a z particle, yields a z particle. This is, of course, an arbitrary prescription. It turns out that there are about 100 different such prescriptions, even considering as different only such which cannot be made to coincide by permutations of the letters. One will assume that a rule, once chosen, remains valid for subsequent generations and that fixing the rule, one obtains the proportions of each color in the course of time by iterating the initial transformation (that is to say, composing it with itself). Mathematically the problem is in two variables only, since if $x + y + z = 1$, it is also true that $x' + y' + z' = 1$, and so on. One can therefore consider the iterates of the transformation in two variables only.

It turns out that, depending on the initial rule, a great variety of cases arises. In some, starting with any initial distribution, i.e., with any point in the plane, the iterates converge to a fixed point, which indicates convergence of the population to a stationary distribution. For other rules, however, one observes convergence, not to an interior point, but to an oscillating system. Paul Stein and the writer have examined a case in four variables and have obtained a periodic limiting cycle of the order of 12. In some other cases, there is a convergence to strange limiting curves. For cubic transformations, still in the plane, one obtains in the limit weird curves as "cyclic" distributions. In some transformations, points seem to converge to a perfect, nowhere dense set S lying on a curve. In the present state of analysis, it is impossible to calculate this curve in any explicit way, or even, for that matter, to establish its existence a priori. Recently, Stein found a case of quadratic transformation of the above kind, in four variables, in which, starting with any interior points, the iterates converge to two skew curves on which they run infinitely, switching from one to the other.

The ergodic properties of the iterates of a linear transformation, e.g., rotations of a circle, are well understood from the theorems of Kronecker and Weyl. For quadratic point transformations, asymptotic behavior remains mysterious; our heuristic work has shown a bewildering array of different cases with strange behavior. It is hoped that some hints toward understanding the properties of nonlinear transformations will eventually come from this experimental material.

In problems of this nature, it is very important to have some auxiliary equipment attached to the computer to display the results of the calculations. An examination of the coordinates of the points printed out or listed on a tape is impractical. If one can display visually the positions of points in a plane, or in three dimensions in projections, or better still, in axonometric view, one gets very quickly an impression of the morphology of the distributions. Such a display mechanism, constructed by J. Richardson in Los Alamos, has shown itself of great use. In the problem mentioned above, the visual display enables one to see readily the iterates of a point traveling on curves or pieces of curves. This display can be repeated often enough so that one has more than fleeting impressions. Furthermore, the collections of points can be photographed (Figures 3 to 5).

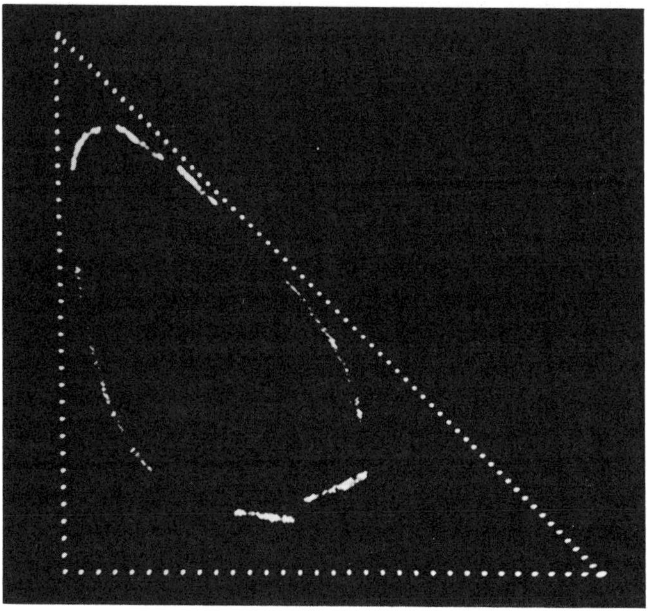

Figure 3. Several hundred iterates of an initial interior point under the transformation

$$x' = z^3 + 3xy^2 + 3yx^2 + 3yz^2 + 3zy^2 + 6xyz$$
$$y' = 3xz^2 + 3zx^2$$
$$z' = x^3 + y^3$$

where $x + y + z = 1$.

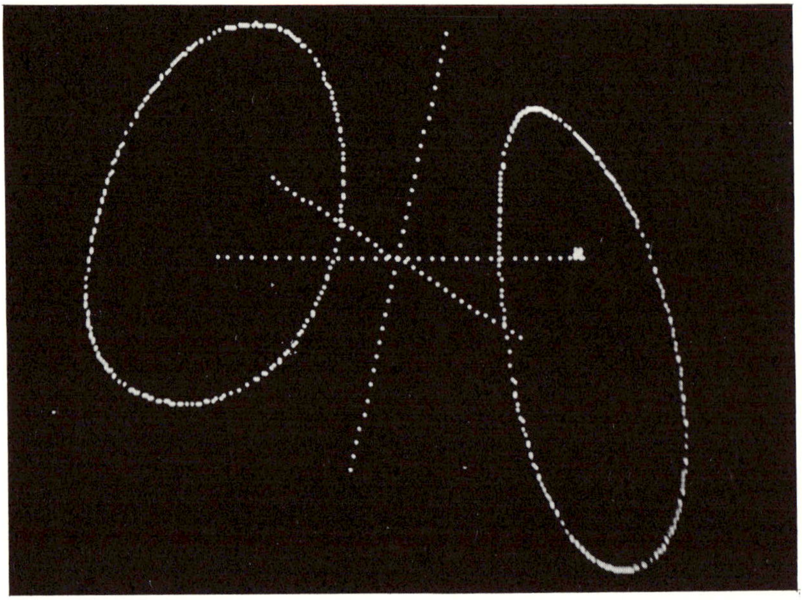

Figure 4. Four hundred iterates of T, shown in two-dimensional projections, with the x, y, and z axes indicated, under the transformation

$$\begin{aligned} x' &= x^2 + y^2 + 2yu \\ y' &= u^2 + 2xu + 2zu \\ z' &= z^2 + 2xy + 2xz \\ u &= 2yz \end{aligned}$$

where $x + y + z + u = 1$.

3

In 1953, the late Enrico Fermi and the writer started a small program of systematic investigations of the long-time, asymptotic behavior of nonlinear physical problems.[2] The first example chosen was a simple one. A string with fixed ends is displaced from equilibrium. It vibrates under the usual law of force, which is linear, but with an added nonlinear term for the force as a function of displacement. The nonlinear term (quadratic, for the first class of problems) is small compared with the predominant linear expression. The purpose of the calculations, which were performed on electronic computers, was to determine whether in the course of time the shape of the string, started initially at time $t = 0$ in a single sine mode, for example, would gradually become more and more complicated, and how the energy, initially all in the first mode, would flow to higher Fourier frequencies. Would there ultimately be a convergence to an equipartition of energy among all the possible frequencies?

One studies the problem on a computer by replacing the continuum of the string by a finite number of points (64 in our case); then the partial differential equation becomes a system of difference equations. The continuous time parameter is also replaced by a discrete number of time cycles. If x_i is the displacement of the ith particle of the string, the equations are of the type

$$\ddot{x}_i = (x_{i+1} + x_{i-1} - 2x_i) + \alpha[(x_{i+1} - x_i)^2 - (x_i - x_{i-1})^2]$$

or

$$\ddot{x}_i = (x_{i+1} + x_{i-1} - 2x_i) + \beta[(x_{i+1} - x_i)^3 - (x_i - x_{i-1})^3]$$

where $i = 1, 2, \ldots, N$, and α and β are coefficients so chosen that, at the maximum displacement which occurs, the nonlinear term amounts, at most, to one-tenth of the force. The problem was studied by following tens of thousands of time cycles or hundreds of the expected periods of the linear vibrations.

The results of the computation were very surprising! Instead of a gradual flow of energy to higher harmonics and a tendency toward

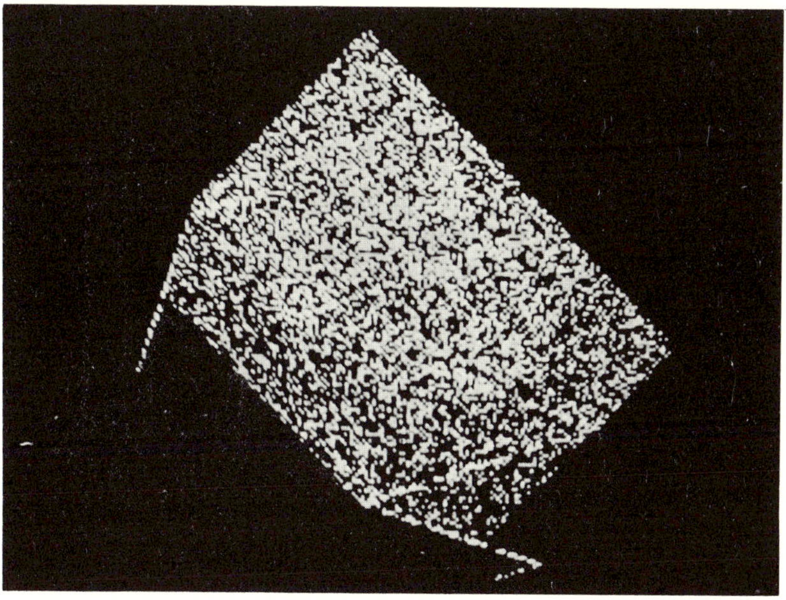

Figure 5. Approximately 9,000 iterates of a point interior to the square, under the transformation

$$x' = f(x, y) \qquad y' = g(x, y)$$

where f and g are functions which equal 0 on the circumference of the square and 1 at the points $(\frac{1}{4}, \frac{1}{4})$ and $(\frac{3}{4}, \frac{3}{4})$, respectively. They each increase linearly from 0 to 1 in the four triangles inside the square; that is, f and g form pyramidal surfaces.

equipartition, something quite different was observed. Only the few low modes acquired any noticeable energy. There was an interplay among those until, after about thirty thousand cycles of computation, the string returned to its original form to within closer than 1 per cent of the total energy (Figure 6). It was suspected, at first, that this behavior might be due to the special, mathematical nature of the force. Similar phenomena were observed, however, when different expressions, e.g., a small cubic term, were substituted. Instead of a gradual sharing of energy among the various frequencies, there was a distribution of energy among a few low frequencies, which permuted among themselves.

In this first experiment, the string returned to within 1 per cent of its original condition. Starting with this configuration, J. Tuck and Mary Menzel continued the computation for much longer time intervals. After another period of this sort (corresponding to several hundred of what would have been the linear operation period), the string returned to within 2 per cent of its original position. The "great periods" were repeated several times. After eight of these large periods, the discrepancy was as much as 6 per cent. But, from then on, the discrepancy

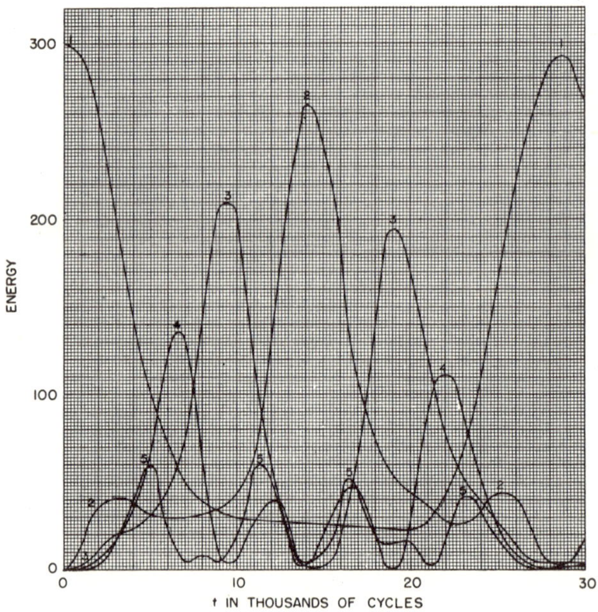

Figure 6. The quantity plotted is the energy (kinetic plus potential) in each of the first five modes; about 30,000 computation cycles are calculated. The units for energy are arbitrary.

began to diminish, and after 16 large periods the string was almost *exactly* in the initial position again; in fact, to within less than the 1 per cent of the original great cycle. This phenomenon reminds one of the cycles of cycles in Hindu mythology.

During the last few years, several mathematicians have examined these heuristic results with a view toward possible explanations through analytical work. Recently, Kruskal and Zabudsky obtained most interesting results on partial differential equations which correspond to the limits of our difference schemas. It appears that such nonlinear equations have solutions which tend to develop discontinuities in the higher derivatives, leading to phenomena of the kind described.

4

The increasing complexity of ideas in even the fundamental theories of physics, in addition to the constantly augmenting combinatorial and analytical complexity of recent technology, has made the computing machine not only welcome, but necessary, as an aid in scientific work. In Los Alamos, during the war, study of the phenomena determining the nature of nuclear explosions involved a treatment of simultaneous problems in neutronics, hydrodynamics, and the behavior of radiation. In order to obtain a quantitative answer, sometimes merely of the yes-or-no kind, enormously elaborate calculations had to be performed. Concepts in basic physics, considered to be elementary or fundamental, have repeatedly turned out to be systems of many interrelated aspects. This tendency, characteristic of recent developments, has led to constantly increasing mathematical sophistication of the simple underlying picture developed during the nineteenth century.

But even in the somewhat older theory, for example, in the theory of general relativity and gravitation, the mathematical difficulty of obtaining particular solutions is very great. In fact, it is not now possible to check the predictions of the general theory without numerical work; to decide between alternative formulations, it is necessary to calculate their consequences. A similar situation prevails in the proposed field theories in nuclear physics.

In still older subjects, e.g., in the field of hydrodynamics, the situation is the same, and methods of the kind described offer a profitable approach. The problem of the onset of turbulence, for example, can be studied by calculating specific cases. These "experiments on paper," using computers, have already been started by many mathematicians and physicists, and will undoubtedly be continued. A number of problems in the mechanics of continua have recently been studied by brute force, so to speak. The fluid is replaced by hundreds or thousands of material points, and forces, such as the gradients of pressure, are simulated by repulsions chosen as a function of the distance between the points.[3] The results of this work give promise, during the next few years, of producing a much better understanding of the dynamics of compressible media. One application of the knowledge thus acquired will be the possibility of calculating more precisely than heretofore possible the phenomena of circulation in the atmosphere, and consequently of predicting the weather. Understanding of the fundamental problems of stability of atmospheric motions may lead, ultimately, to some control of air masses, perhaps even with moderate expenditures of energy. It should be noted again that, for problems of this kind, improved display mechanisms and easier access to the computer, for controlling the course of calculation, are of the utmost importance.

Space does not permit a detailed description of the interesting applications of computing machines in problems of astronomy. Work on stellar models and on the evolution of stars has benefited greatly by the use of computers. The work of Henyey, Schwarzschild, Hoyle, and others has been of the greatest interest and value in cosmogonic theories. With the new high-speed computers now in operation, it is possible to tackle the problem of the equilibrium and development of rotating stars. These machines also make possible advances toward the solution of problems in statistical mechanics and quantum physical chemistry.

5

It may appear that the promise of the pragmatic approach has been emphasized largely in the more abstract parts of mathematics. It is well to remember that some branches of science originated as the study of isolated curiosities or "mathematical amusements." But these soon acquired importance, beyond mathematics, in physics or statistical mechanics, for example. The theory of probability and the field of combinatorial analysis (an ill-defined area of mathematics dealing with the properties of arrangements of objects) originated in problems of chance or mathematical games. Somehow, perhaps through biological instinct or the "wisdom of the race," these theoretical activities acquire a vital, practical significance for the species as a whole. An example that comes to mind is the recent application of combinatorial mathematics to the rapidly developing field of molecular biology. A most exciting and important work is now in progress on the understanding of the code in the DNA molecules. The study of the sequence of the molecules in a linear structure, which determines the geometric properties of protein molecules, is now possible because of the availability of electronic computers. By sophisticated mathematical analysis, Kendrew in England determined the relative positions of the atoms constituting the myoglobin molecule, work which could be done only because of the availability of mechanical means of performing the necessary calculations.

In the theory of specialization of cells and the structure and growth of their groupings, mathematical models, examined by the aid of computers, will no doubt be of great use. One class of problems will deal with the diffusion process of a great many particles of different types, among which short-range binary forces tend to organize diverse subsystems.

Going further, it is perhaps permissible to speculate on a still different role of the computing machines. It is conceivable that, through experimentation with computers, new types of intuition may be acquired by mathematicians; these may include "palpable" experience with objects in more than three dimensions and greater familiarity with new topological and kinematical situations. It is only too easy to underestimate the heuristic value of such approaches. But some anthropologists have seriously advanced the thesis that it was the development of the opposable thumb in the hand, and the ensuing use of the hand and tools, that helped to a considerable extent to mold and develop the human brain itself. It is foolhardy to prophesy where the development of such new tools will lead us in the future, but it seems quite clear that the progress of technology, rapid and alarming as it is, may be accelerated still further.

A great deal of preliminary speculation and some work have been devoted to what is called the analogy between the structure and connections in the computers, on the one hand, and the neural connections in the nervous system and in the brain, on the other. This is an enormous subject, about which nothing very definite can be said as yet. But even if we discover, by looking at the apparent analogies, what the brain is not, a great deal will be accomplished.

It is inevitable that the increasing availability and use of computers will bring about modifications in some of the curricula of instruction in mathematics and other sciences. This is already apparent on the undergraduate level and will lead to considerable changes in educational programs. In the near future, the young scientist will do his work and develop his ideas with the aid of novel and valuable tools, among the most important of which will be electronic computers.

REFERENCES

1. P. Stein and S. Ulam, Quadratic Transformations, Part I, *Los Alamos Sci. Lab. Rept.* LA-2305, Office of Technical Services, U.S. Department of Commerce.
2. See, for example, S. Ulam, "A Collection of Mathematical Problems," Interscience, New York, 1960.
3. F. Harlow et al., The Particle and Force Computing Method for Fluid Dynamics, *Los Alamos Sci. Lab. Rept.* LAMS-2567, 1961.

Edward G. Bowen, Chief, Division of Radio Physics, The Commonwealth Scientific and Industrial Research Organization, Sydney, Australia, was an original member of the English radar development team in 1935. He holds the Medal for Freedom of the United States government for his contributions to radar development in this country.

CHAPTER 6

RADIO ASTRONOMY AND RADIO TELESCOPES

1

Since the dawn of history, man has gazed at the stars and wondered about the mystery and the immensity of the universe. He did this first with his naked eye, then with the help of crude lenses no better than a burning glass. Over the years, his understanding of the universe has improved and his instruments have become more powerful, culminating in the outstanding discoveries that were made during the first half of this century with the great 100- and 200-in. optical telescopes of Mt. Wilson and Mt. Palomar.

Only in the past thirty years have radio telescopes appeared, and only in the last fifteen years has the study of radio astronomy received really serious attention. But already giant instruments exist in the Northern and Southern Hemispheres that are capable of penetrating into the universe at least as far as, and possibly farther than, the best optical instruments.

Radio astronomy began in the 1930s with an observation by Jansky,[1] who was then involved in a study of the static which affected trans-oceanic radiotelephone circuits. Using a rotating, directional antenna system operating at a wavelength of 14.6 meters, he found that after all other noise sources had been accounted for, there was still a faint residual hiss in his receiver. He concluded that these electromagnetic signals originated in sources beyond the earth, probably from the direction of the Milky Way.

This observation was confirmed and extended by Grote Reber,[2] a radio amateur, who built an elementary form of radio telescope in his back yard. He measured the distribution of cosmic noise at much higher frequencies and with a sharper antenna beam.

Unfortunately, the significance of these pioneer observations was not recognized at the time. It was not until the end of World War II that interest was revived and studies of radio astronomy were carried out in England, Holland, the United States, and Australia.

Since light and radio waves are essentially the same except for length, it is to be expected that the design and use of optical and parabolic radio telescopes present similar problems. Thus the resolving power of both types is described by the same relation,

$$\theta \sim \frac{\lambda}{D}$$

where θ = angular beam width or limit of resolution, λ = wavelength, and D = aperture (diameter) of the instrument. But radio waves are so much longer than light waves that, to compare in resolving power with the best optical telescopes, a radio telescope would need an antenna many miles in diameter. It may be added, however, that the position of a radio source that is much brighter than its surroundings can be located with greater angular precision than is indicated by the resolving power of the radio telescope.

The difference in wavelength has another consequence. Interstellar dust, which prevents optical observation of many parts of the universe, is transparent to radio waves. Furthermore, the electromagnetic band of optical frequencies is 5 octaves wide, whereas radio astronomy now has available a band 12 octaves wide. Short radio waves, of the order of a few millimeters in length, are absorbed by the oxygen and water vapor of the atmosphere, while waves longer than a few decameters are reflected back into space by the ionosphere.

The resolving power of the most precise radio telescopes, operating at microwave frequencies, barely approaches that of the human eye in the range of visible light. Hence the radio telescope pictures an indistinct universe, dustless, but populated largely by clouds of gas, with the more intense radio stars appearing as blurred images of about the size of the sun or the moon.

Radio telescopes now exist in a bewildering array of types, but only the simplest, and perhaps the most elegant, of these will be considered, namely, the steerable reflector type (Figure 1). This usually consists of a parabolic surface which collects the incoming radio waves and brings them to a focus in the focal plane. The signals are then detected by a very sensitive receiver, and the output of this receiver is led to a pen recorder, which plots the signal as a function of time.

If a radio star or an extraterrestrial source of radio waves passes through the beam of the radio telescope, the signal increases rapidly as the source enters the beam, rises to a maximum when it is in the center of the beam, and decreases to the original background level as it goes out of the beam (Figure 2). It is from a study of the characteristics of this signal that our knowledge of the physical characteristics of the source is gained. For example, the position of the source is determined from the position and time of the maximum, and information about the size of the source, from the width of the trace. From observa-

Figure 1. Steerable parabolic radio telescope, 210 ft in diameter, nearing completion, Parkes, New South Wales, Australia.

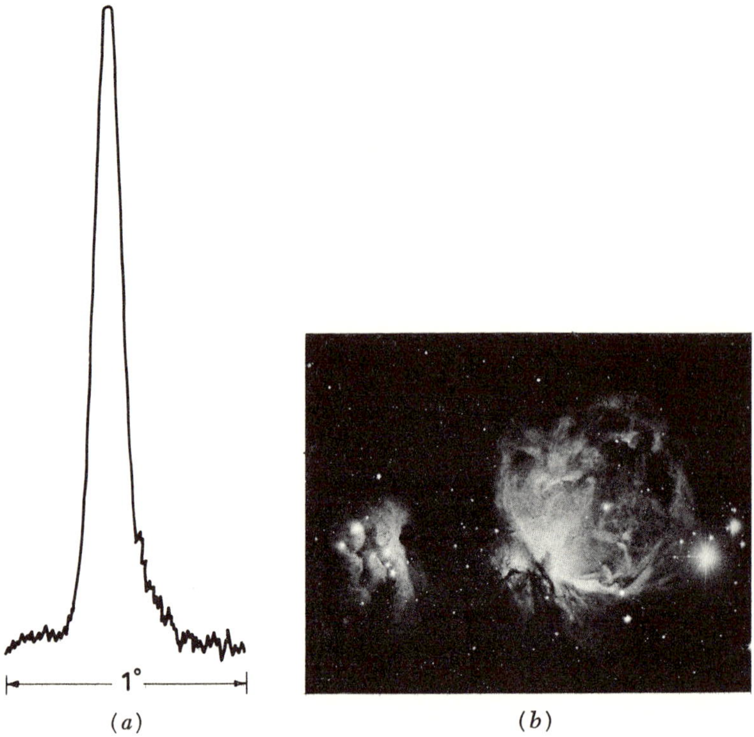

Figure 2. (*a*) Recording of the Orion Nebula (M42) as it passes through the aerial beam of the Australian 210-ft radio telescope. (*b*) Orion Nebula. (Courtesy of Lick Observatory.)

tions at a variety of frequencies, it is possible to determine the distribution of intensity over the radio spectrum. In this way, something can be learned about the physical processes taking place in the source, for example, whether the radio signals are generated by plasma-type oscillations or by synchrotron-type, or else by simple thermal radiation at radio frequency. Each of these gives rise to different spectral intensity distributions, and it is often possible to distinguish clearly among them.

The study of signals from distant sources gives new insights into the nature of the universe, and in the past fifteen years a great mass of new information has accumulated. What has been learned does not supplant what was already known from optical astronomy: on the contrary, radio astronomy can now be said to complement optical astronomy, as an important and highly productive method of gaining knowledge about the most distant parts of the universe.

2

The sun is one of the most interesting radio sources. It consists of the visible disk, the photosphere, which is at a temperature of about 6000°K; the chromosphere, a thin layer at a temperature of about 10,000°K; and the corona, the sun's atmosphere, extending several solar diameters.

The quiet sun generates radio noise signals at a comparatively low level of intensity. These signals come, not from the photosphere, but from the corona, which is not normally visible, but can be seen clearly at times of total eclipse (Figure 3*b*). The corona, which is both extremely hot and highly ionized, radiates in the thermal mode. A typical plot of the intensity of the radio signals obtained from the sun at a wavelength of 20 cm is shown in Figure 3*a*.

If the radio receiver is operated at a comparatively long wavelength, for example, 10 meters, the signals received must originate in the outermost edge, since considerations based on electron density show that 10-meter signals cannot be propagated outward from the depths of the corona. By decreasing the wavelength to which the receiver is tuned, signals are received from intermediate levels of the corona, down to very nearly the sun's surface.

The radio telescope and receiver detecting the signals act like an optical pyrometer, so that it is possible to measure the temperature gradient in the corona. It is found that the temperature increases outward, from 10,000°K at the surface of the sun to almost a million degrees at the edge of the corona.

The disturbed sun is subject to solar storms which make themselves visible as sunspots or as vast flare outbreaks. The latter are accompanied by explosive outbursts of material from the surface.

Radio studies have shown that following a prominent flare the following sequence of events may occur.

1. Almost simultaneously with the appearance of the flare, a stream of very fast particles is thrown up through the corona at a velocity which may be one-third or one-half the velocity of light. These particles do not themselves produce radio waves, but plasma-type oscillations are generated as they tear their way out through the sun's atmosphere. They may be considered in some ways as the electromagnetic equivalent of the acoustic shock wave generated by an aircraft flying at supersonic velocity.

These fast particles were first identified, and their velocities measured, by means of the radio waves which accompany their passage through the sun's atmosphere. They have since been shown to be asso-

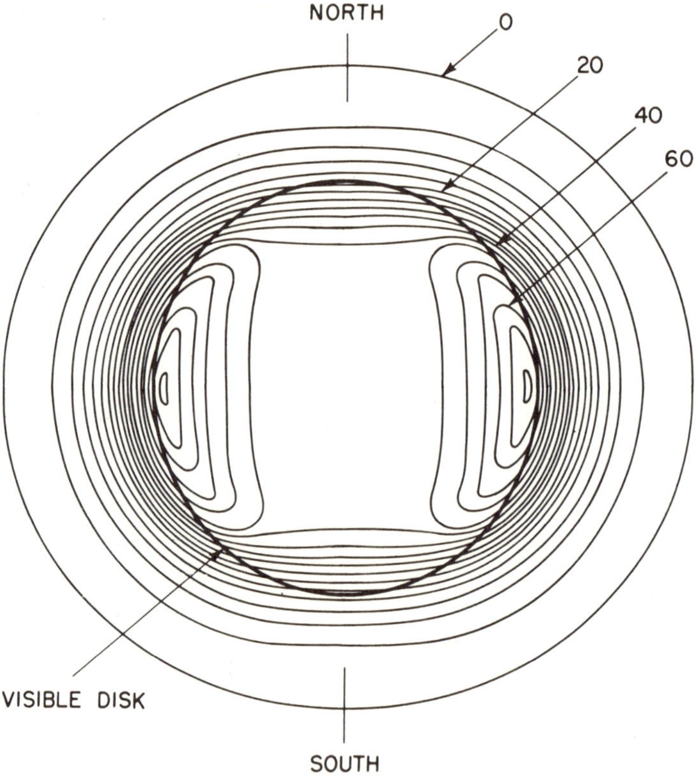

(*a*)

(*b*)

Figure 3. (*a*) Radio contours of the quiet sun at a wavelength of 20 cm. (*b*) Solar corona photographed during a total eclipse of the sun.

ciated with an increase in cosmic-ray intensity occurring on the earth 30 to 60 min after the flare event on the sun.

2. A few minutes after the high-speed particles are ejected, a second stream of slower particles appears, traveling at the comparatively low velocity of 500 to 1,000 km/sec. These particles generate radio waves in much the same way as the high-speed particles, and their passage may be traced out to about one solar diameter from the sun's surface.

When these particles reach the earth, about 24 hr after leaving the sun, they are responsible for events the occurrence of which has been known for some time. For example, they cause magnetic storms, give rise to auroras, and create disturbances in the ionosphere which interfere with long-distance radio communications.

These particles were known to exist before the radio observations were made, but radio measurements provide an independent determination of their velocity and, probably more important, offer a means of locating the events that give rise to them, even if the sun is obscured by cloud.

3. Lastly, for some hours after the sequence of events just described, a large and more diffuse source of radio waves drifts slowly up through the corona. On some occasions, it has been tracked as far as five solar diameters from the sun. These signals are known to be generated by the synchrotron process and are clearly related to the magnetic fields which exist around these disturbed regions on the sun.

The entire process may be likened in a crude way to the firing of a cannon in the field. First there is the flash (in this case, propagated with the velocity of light) which might be compared to the fast particles; then comes the projectile, which corresponds to the 1,000-km/sec particles; and finally, the puff of smoke which drifts slowly away.

Radio waves are emitted by the planets of our solar system, but there is as yet relatively little information about their origin. In the regions beyond the solar system are the so-called radio stars, discrete sources of radio waves found in profusion in most of the sky.

The positions of many thousands of these have now been plotted. It is a remarkable fact that in general only a relatively small number, perhaps 10 per cent, correspond in position to known visible objects. When a visible object does correspond to a radio source, it nearly always turns out to be a highly unusual one in many respects.

One example is the well-known Crab Nebula (Figure 4*a*), first recognized as a radio source by John Bolton in 1949,[3] which was seen by Chinese astronomers as a supernova in the year 1054. When first ob-

(*a*)

(*b*)

Figure 4. Some radio sources that have been identified. (*a*) The Crab Nebula the "radio galaxy" Virgo A (NGC 4486), with the unique jet extending from its radio source is located within apparently blank area indicated; (*d*) the Great Wilson–Palomar Observatories.)

(*c*)

(*d*)

(M1), the remains of a supernova explosion which occurred in A.D. 1054; (*b*) nucleus; (*c*) star field in the constellation Cygnus: the historic "colliding galaxies" Nebula in Andromeda (M31), the closest of the spiral galaxies. (Courtesy of Mt.

served, it was bright enough to be seen in daylight, but it then decreased in intensity and slowly increased in size, until today it is no longer visible to the unaided eye, but covers an area in the sky larger than our own sun. This is truly an explosion on a cosmic scale; it is not surprising that this object is among the three strongest radio sources in the whole sky.

Another radio source of great interest is that in the constellation Virgo (NGC 4486), which, when observed in white light, shows only a diffuse circular nebulosity. When photographed in blue light, however, it exhibits a remarkable jetlike structure, reminiscent of the jets and prominences thrown up from the surface of the sun, but on a gigantic scale (Figure 4*b*).

The first discrete radio source to be discovered was that in the constellation Cygnus (Figure 4*c*). For many years, no optical object could be detected at the source, but an intensive search by Baade and Minkowski [4] with the 200-in. telescope at Mt. Palomar finally showed a diffuse nebulosity, which, after further study, they considered to represent two spiral nebulae in collision. This possibility was highly attractive at the time, but it has recently been questioned, and the resulting doubt is likely to be resolved only by further observations of an optical nature.

But the existence of the source in Cygnus raises another interesting and challenging point. This source is the second strongest in the whole sky, matching in intensity the signals from the quiet sun. Nevertheless, it is at the very limit of the range of the 200-in. optical telescope. Might it be, therefore, that the thousands of radio sources for which no optical equivalent can be found are *beyond* the range of the largest optical telescopes?

Returning from the outermost reaches of the universe, we consider now our own star system, the localized concentration of stars, aptly known as the Milky Way, which is seen like a band across the sky. Extending across both the southern and the northern skies, it provides visual evidence of the fact that we lie within a vast saucer-shaped assembly of stars.

The broad characteristics of this system have been known for some time from optical observations. For example, its dimensions are such that it takes a light wave or a radio wave about 100,000 years to travel from one side of the system to the other, but only 10,000 years to travel across the thin dimension. It is also known that the solar system, far from being at the center of things, is located well out toward the edge.

Finally, the whole system rotates, like coffee in a cup, with an angular velocity that is highest near the center and decreases toward the edges.

However, until the advent of radio astronomy, little was known about how our star system would look to an observer elsewhere in the universe. The situation was changed in a dramatic way by a discovery of Ewen and Purcell,[5] which is perhaps the most important single observation made in the young science of radio astronomy.

The radio signals so far considered are all random noise signals, the energy of which is distributed widely over the radio spectrum. In 1951, however, Ewen and Purcell detected a signal coming from the direction of the Milky Way, a line frequency of 1,420 Mcps, which must come from hydrogen atoms.

This observation was important for a number of reasons. A great part of the mass of a galaxy is due to the interstellar gas, which is mainly hydrogen. Because of the low temperature of interstellar space, most of the hydrogen is in the ground state and, as a consequence, difficult to detect with optical instruments. Study of this radio line is therefore a valuable tool in obtaining information about the nature and physical state of the emitting region. Furthermore, since it is a line frequency, it is possible to measure Doppler shifts, which indicate whether the hydrogen is approaching or receding.

By observation of the hydrogen-line radiation from the plane of the Milky Way, it has been possible to map the distribution of hydrogen in our galactic system, with a result which is both intriguing and gratifying. The system turns out to be a vast spiral galaxy (Figure 5), similar to the millions of other galaxies which float in space like so many island universes. There is nothing unusual or peculiar about the structure of our system.

In radio astronomy, radiations emitted by the bodies under study are observed and measured. In the closely related field of radar astronomy, signals are transmitted from the earth, and the reflected signals are then studied. Part of the transmitted pulse is scattered, and part is absorbed; the power of that portion of the scattered part that is reflected back to the earth is so small that special techniques are required to detect and recognize the echo. Nevertheless, echoes of this kind have been received from the sun, the moon, and Venus; several other planets will probably be reached during the next few years.

3

The story of the development of radio telescopes begins with Reber's original instrument in Illinois, 31 ft in diameter (Figure 6). This was

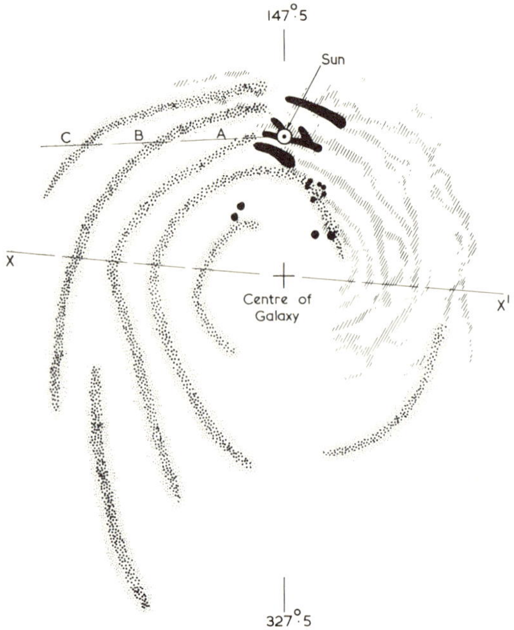

Figure 5. A composite map of the spiral structure of our galaxy, derived from observations made at Sydney (left) and Leyden (right).

recently reerected, as a memorial to his pioneer efforts, at the National Radio Astronomy Observatory at Greenbank in West Virginia. Then came the 50-ft instrument at the Naval Research Station in Anacostia (Figure 7), followed by 60-ft instruments, such as the one at the Harvard Observatory (Figure 8). Next came the 85-ft dishes (Figure 9) used both for satellite tracking and for radio astronomy, and finally, the 90-ft telescopes at the Bishop Radio Observatory of the California Institute of Technology (Figure 10).

Telescopes in this size range now exist in comparatively large numbers in many countries of the world and are likely to be the basic tools of the radio astronomer for many years to come.

The largest steerable instrument in existence, the pioneer of the real giants, is the 250-ft paraboloid at Jodrell Bank in Manchester, England (Figure 11). In the past few years, it has been used to perform outstanding work in the tracking of space vehicles. The experience gained in its operation has provided inspiration and guidance to other astronomers who are interested in the construction of large instruments of this kind.

About six years ago, it became clear to Australian radio astronomers that most of the investigations which could be done with relatively simple equipment had been carried out and that a giant steerable telescope was needed for further work. Consideration was therefore given to the requirements for such an instrument. These may be summarized in general terms.

1. The telescope should be as large as possible, subject only to the limitations imposed by the mechanical properties of the material used in its construction and by the amount of money available. This is one of the few cases where sheer size has merits: first, because the power gain of the antenna, and hence the distance to which it can penetrate into the universe, increases with increasing size; and second, because the resolution, which should be as great as possible, increases linearly with size. It appeared from these considerations that a telescope in the size range 200 to 250 ft was both desirable and possible, but that to

Figure 6. Reber's original steerable parabolic telescope, erected at his home in Wheaton, Illinois. (Courtesy of *Astrophysical Journal.*)

Figure 7. The 50-ft reflector of the United States Naval Research Laboratory, Anacostia. (Courtesy of United States Naval Research Laboratory.)

aim for a diameter greater than 250 ft might invite both engineering and financial difficulties.

2. The instrument should have superlative performance for hydrogen-line observations at a wavelength of 21 cm. Therefore the reflecting surface of the dish should conform to a true paraboloid to an accuracy of at least $\pm\frac{1}{2}$ in., and the focal point should be held to the same accuracy. These characteristics would have to be maintained in the face of gravity loads, wind loads, and variations of temperature.

3. To match the highly accurate surface, the ability to point the telescope to any part of the sky with an accuracy of 1 minute of arc would be required. It should be emphasized that this is not merely the precision with which the stationary telescope could be pointed, but the accuracy with which a radio source could be tracked across the sky.

Just as important as the statement of requirements for an instrument of this kind is a knowledge of which requirements may be relaxed when engineering difficulties start to mount. In this respect, a number of important concessions were made. For example, in radio astronomy there is no real need to operate in a high wind. The highest surface accuracy and the highest pointing accuracy were therefore specified only up to wind speeds of 10 mph; a reduction in accuracy by a factor of 2 was allowed up to wind speeds of 20 mph.

It has turned out that the performance of the finished instrument is much better than this and, in fact, gives full accuracy at wind velocities above 20 mph. But knowing that the specifications could be relaxed in this way was an important contributing factor leading to an economical design.

Similarly, there is no great need to operate the telescope down to the horizon, where noise from terrestrial sources begins to interfere and where refraction effects become severe. The savings in this regard are mainly financial, since the cost of the structure increases very rapidly as the dish is required to point closer and closer to the horizon.

When the requirements had been determined, the next steps involved

Figure 8. The 60-ft equatorially mounted reflector at Harvard College Observatory. (Courtesy of Harvard College Observatory.)

Figure 9. The 85-ft equatorially mounted radio telescope of the California Institute of Technology–Jet Propulsion Laboratory at Goldstone, California. (Courtesy of California Institute of Technology.)

a design study and then the preparation of drawings and specifications from which a telescope could be built.

An important outcome of the design study was the calculation of a size-cost curve for telescopes in the 200- to 250-ft size range. It transpired that the cost would vary approximately as the diameter to the 2.7 power; i.e., doubling the size of the instrument would increase the cost by a factor of almost 8. But it was also obvious that this curve was valid only over a limited range of sizes and that as the size increased still further, the engineering difficulties would multiply and the cost might approach the gross national budget.

The prudent course, therefore, was not to press the size too far, but to settle for the largest telescope that could be built for a reasonable cost and without excessive engineering difficulty. The size finally chosen for the Australian telescope on the basis of these considerations was 210 ft in diameter.

Figure 10. The California Institute of Technology radio telescopes at the Owens Valley Radio Observatory. (Courtesy of California Institute of Technology.)

Figure 11. The 250-ft steerable paraboloid at Jodrell Bank, Manchester, England. (Courtesy of University of Manchester.)

The structure of the Australian telescope rests on a reinforced-concrete tower, supporting a steel track 40 ft in diameter on which the telescope rotates about a vertical axis (Figure 12). On this track is carried a massive turret (Figure 13), at the top of which there is a cylindrical drum about 20 ft deep and 20 ft in diameter. This forms the hub of the main reflector structure, which is then built up from this hub by conventional rib-and-ring girder construction.

This framework carries the mesh panels which reflect the radio waves to the focal point (Figure 14). The panels are mounted by means of adjustable bolts, which may be used subsequently to trim the surface to the utmost precision.

At the focal point is a radio cabin, an 8-ft cube, mounted on a large tripod. The floor of this cabin carries the small pickup antennas, which receive the signals reflected from the parabolic surface; it also houses the highly sensitive preamplifiers, which feed incoming signals through flexible connections to a radio receiving room in the base of the structure.

The entire structure is of steel. The weight on the azimuth track approaches 1,000 tons; the full load on the foundations is about 1,500 tons.

To achieve the desired tracking accuracy of 1 minute of arc, a device

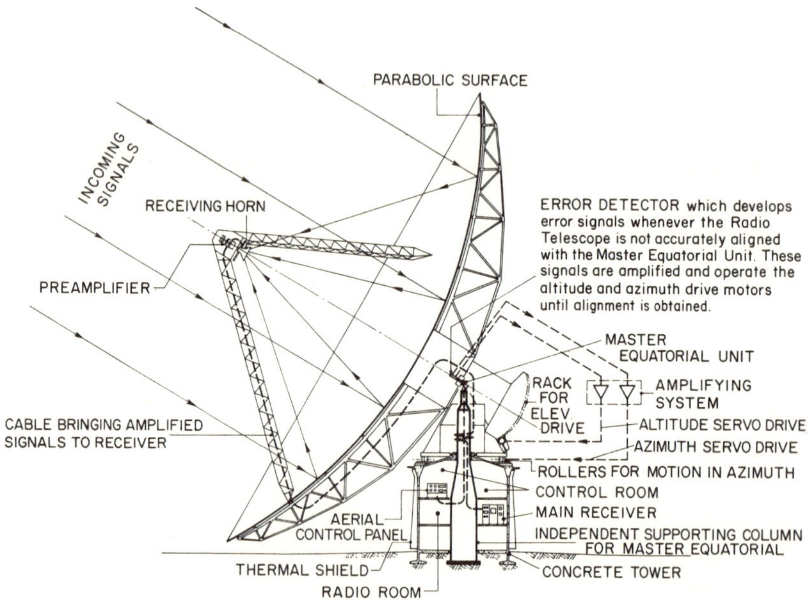

Figure 12. Diagrammatic representation of the Australian 210-ft radio telescope.

Figure 13. The Australian telescope under construction: turret and hub, on which the ribs are being assembled.

Figure 14. The inner section of the reflector, to a diameter of 54 ft, is covered with steel plate, and the remainder with panels of $\frac{5}{16}$-in. mesh of high-tensile-steel wire.

invented by Dr. Barnes Wallis has been adopted. A master control unit is installed at the intersection of the two axes of rotation of the telescope (Figure 15). This unit resembles the equatorial mount on which a conventional optical telescope is carried and is capable of a high order of accuracy. A light beam is directed along the hub of the radio telescope to the master equatorial unit, and any error in alignment between the two is fed to the servomotors driving the main telescope. In this way, the radio telescope becomes a slave to the smaller, but very precise, master system and follows the correct motion to a high order of accuracy (Figure 16).

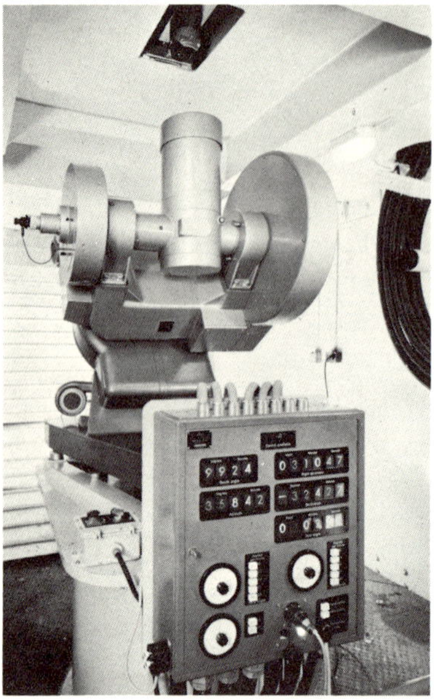

Figure 15. Master control unit of the Australian radio telescope.

Actual construction occupied approximately two years, the first of which was spent on the fabrication of parts, and the second in erecting the instrument at a site near Parkes in New South Wales, which was specially selected for its low electrical noise level and low average wind. The instrument was finally brought into commission by His Excellency, Lord de L'Isle, Governor General of Australia, at an inaugural ceremony on 31 October 1961 (Figure 17).

The assistance of the Carnegie Corporation and the Rockefeller Foundation is acknowledged with gratitude. These two institutions contributed nearly half the cost of the instrument, the remainder coming from private sources in Australia and from the Commonwealth Government.

Although adjustments to the control system were still in process, the first radio observations were made in November 1961, with receivers operating on wavelengths of 10, 20, 21, and 75 cm, respectively. Preliminary results give reason to believe that the performance of the telescope is better than originally planned. It is apparent, for example, that the errors in the reflecting surface do not exceed $\pm\frac{1}{4}$ in. under most operating conditions, whereas the initial specification called for a maxi-

Figure 16. Control desk from which the telescope is operated.

mum error of $\pm\frac{1}{2}$ in. It further appears that there is no deterioration of performance up to wind speeds of 25 mph.

At the time of writing, a final check of the overall tracking accuracy of the telescope has not been made, but the individual parts of the complex control system are operating within specifications. It is expected that tests of the overall system now in progress will indicate that the required tracking accuracy has been achieved.

One of the first records was obtained from a radio source in Orion on a wavelength of 10 cm (Figure 2). It shows that at this wavelength, the telescope has a beam width of 7.5 minutes of arc, compared with the figure of 6.5 minutes of arc expected on theoretical grounds. This performance is vastly superior to any previously achieved in this range of wavelengths.

At a wavelength of 75 cm, approximately three times as many radio sources as previously observed on this wavelength are detected, with one source appearing at average intervals of 10 beam widths.

4

The Australian 210-ft radio telescope will be used predominantly for fundamental studies in the field of radio astronomy.

During the first few months of operation, when the performance of the instrument is still being explored and observers are gaining experience in its use, a series of short observing programs will be carried out to survey a few specially selected regions of the sky.

Figure 17. The Australian 210-ft telescope at minimum elevation, 30°.

The first massive program will be directed toward a hydrogen-line survey of the spiral structure of the Milky Way, carried out in far greater detail than has hitherto been possible and extending to considerably greater distances. The next step will be a similar study of the nearest neighbors outside our own galactic system, the Clouds of Magellan, which probably constitute a galactic system in the early stages of formation.

But perhaps the most challenging and exciting task will be to probe, at distances measured in thousands of millions of light years, the most remote regions of the universe, which have never previously been explored by man.

In penetrating to these great distances, radio telescopes will receive signals which had their origin many thousands of millions of years ago, thus making it possible to look back through the history of the universe, and perhaps to gain insight into how it began and how it has evolved. There have been many theories and many speculations about the nature and origin of the universe, but exceedingly few facts. There are many more questions than there are answers. We do not know whether the universe is bounded or extends an infinite distance in every direction; whether it had a beginning and will have an end, or whether it will exist forever.

Observations made with radio telescopes will undoubtedly provide answers to some of the questions and may in time reveal ultimate truths about the origin and structure of the universe.

REFERENCES

1. Karl G. Jansky, Directional Studies of Atmospherics at High Frequencies, *Proc. IRE*, vol. 20, pp. 1920–1932, December, 1932.
   Karl G. Jansky, Radio Waves from Outside the Solar System, *Nature*, vol. 132, p. 66, July 8, 1933.
   Karl G. Jansky, Electrical Disturbances of Extraterrestrial Origin, *Proc. IRE*, vol. 21, pp. 1387–1398, October, 1933.
2. Grote Reber, Cosmic Static, *Astrophys. J.*, vol. 91, pp. 621–624, June, 1940.
   Grote Reber, Cosmic Static, *Proc. IRE*, vol. 30, pp. 367–378, August, 1942.
3. J. G. Bolton and G. J. Stanley, *Australian J. Sci. Research*, vol. 2A, p. 139, 1949.
4. W. Baade and R. Minkowski, Identification of the Radio Sources in Cassiopeia, Cygnus A, and Puppis A, *Astrophys. J.*, vol. 119, pp. 206–214, January, 1954.
   W. Baade and R. Minkowski, On the Identification of Radio Sources, *Astrophys. J.*, vol. 119, pp. 215–231, January, 1954.
5. H. I. Ewen and E. M. Purcell, Radiation from Galactic Hydrogen at 1420 mc/sec, *Nature*, vol. 168, pp. 356–358, Sept. 1, 1951.

William Shockley, Director, Shockley Transistor, Unit of Clevite Transistor, Palo Alto, California, is the inventor of the junction transistor. He is co-winner of a Nobel Prize for fundamental contributions to transistor physics.

CHAPTER 7

TRANSISTORS†

1

The transistor promises to play a leading role in one of the most important applications of the amplifying principle in man's efforts to control his environment. The concept of amplification is so familiar in the world of electronics that its broad significance is often overlooked.

Fundamentally, amplification is the control of the flow of energy from a source to the point of utilization. In a simple example (Figure 1), the source is the energy of water in a high tower. The relatively small energy input needed to turn the valve controls the output of large amounts of energy. Thus the valve is an amplifier. In fact, in some countries vacuum tubes are called valves and transistors are known as semiconductor valves.

In his conquest of nature, man has used the amplification principle in many ways. He controls the heat output from a pile of burning wood by controlling the position of the sticks of wood. The use of the beast of burden is another application of the same principle. An ox is supplied with energy, in the form of food, which is transformed into chemical energy in its muscles. By pulling on a ring in its nose, one controls the flow of the energy in the ox's muscles; the animal pulls a much heavier load than can be moved by the energy involved in pulling the ring. In

† The presentation of this lecture and its inclusion in this volume were made possible by special arrangements organized by Dr. Carl F. J. Overhage. The time lost by the author due to hospitalization following an automobile accident made it impractical to prepare a manuscript. A stenographic record of the spoken lecture was transmuted into a form and style suitable for printing by Drs. Harry C. Gatos and Robert H. Rediker of Lincoln Laboratory, and the nominal author and the readers are indebted to them for its presence in this volume.

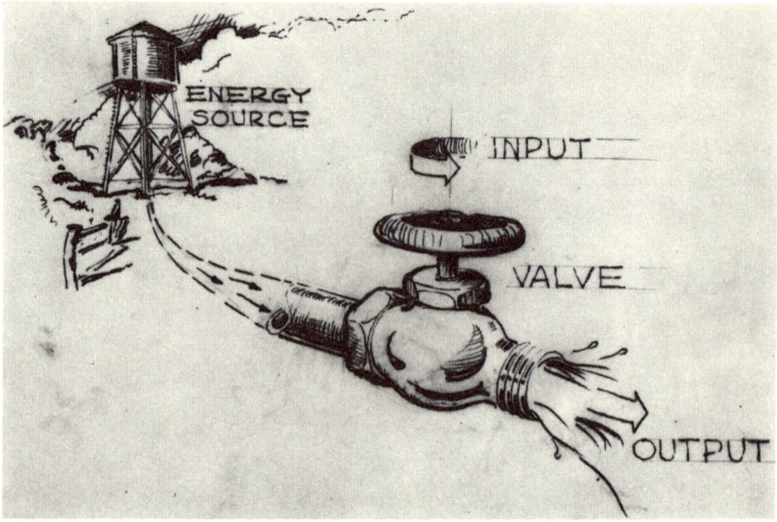

Figure 1. Amplification: control of a flow of energy.

sailing a boat, the wind is the source of energy. By applying a small amount of work in handling the sheets, one controls a very large force which pushes the boat through the water.

Applications of the principle of amplification may be traced in the history of technology, from the early uses of gunpowder and the steam engine to the complex machines of the twentieth century. The most extreme contrast between input and output of energy is found in an amplifying system which is the cause of universal anxiety: the push buttons controlling the detonation of megaton bombs.

As an amplifier, the transistor controls the flow of electrical energy. Despite the difference in time scales, a suggestive analogy may be drawn between the practical applications of electricity in the past century and the corresponding stages in the history of social development during hundreds of centuries (Figure 2). As one of man's earliest technologies involved fire, so one of the first uses of electricity was in the production of heat and light. Similarly, as men discovered how to amplify the energy in their own muscles through the use of beasts of burden and through mechanical engines, so the next application of electrical energy was in electric motors. Then, granting the difficulty of dating the origin of art, music, and writing, we may nevertheless see a parallel between the development of these symbolic forms of expression and the utilization of electricity for radio, television, and communications in general. Finally, science in man's history can be compared with the computer in the history of electricity. The motor is able to do more work than man's unaided muscles; the computer can carry out more calculations faster than the human brain. This is not to say, of course, that a computer is capable of creative thinking, any more than a motor-driven stonecutter can create a statue.

As the motor became cheaper and more compact, it found many more uses. When the equivalent of the motor was a huge steam engine, it could scarcely be used in the kitchen sink for garbage disposal; now that it can be made really small, it can be mounted on a toothbrush. Because of this reduction in size, the motor has an increasingly greater effect on the way we live. Similarly, as the computer becomes smaller, cheaper, and more easily produced, it will undoubtedly exert a greater influence on ordinary life.

APPLICATIONS OF ELECTRICITY	HISTORY OF TECHNOLOGY
LIGHT	FIRE
MOTOR	OX, ENGINE
COMMUNICATIONS	ART, MUSIC
RADIO, TV	WRITING
COMPUTER	SCIENCE

Figure 2. Electricity and the history of technology.

In this process the transistor is particularly important. In one decade, transistors have already reduced the scale of computer components and computers themselves (Figure 3). Furthermore, transistors do not have filaments and therefore do not require as much power to run as do vacuum tubes. Their small size, together with the fact that they can operate efficiently at the low levels of power supplied by small batteries, makes possible their use in miniature devices, such as pocket radios and hearing aids.

These characteristics of the transistor determine its present role and suggest its future significance in electronics technology.

2

The transistor is made of a semiconductor; the principles of its operation are based on the properties of semiconductors. Semiconductors fall between insulators and conductors, being more like insulators than conductors in electrical characteristics. Germanium and silicon, which are the most common semiconductors of which transistors are made, have the well-known crystalline structure of diamond (Figure 4).

In this cubic structure, each atom is symmetrically surrounded by four other atoms. These four atoms are situated in alternate corners of a cube, which is equivalent to occupying the four corners of a regular tetrahedron. An interesting geometrical property is associated with this arrangement: it is the only way in which four balls can be placed

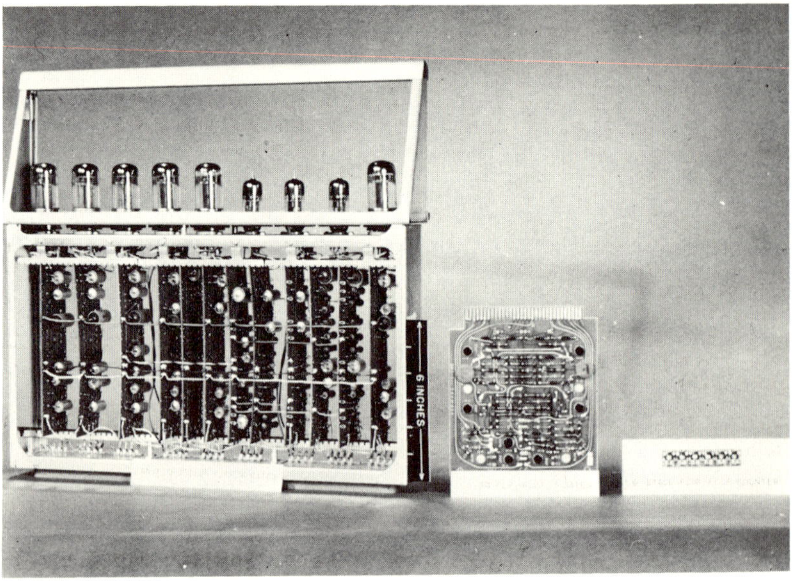

Figure 3. Three generations of the same computer circuit, using vacuum tubes (left) and transistors (right).

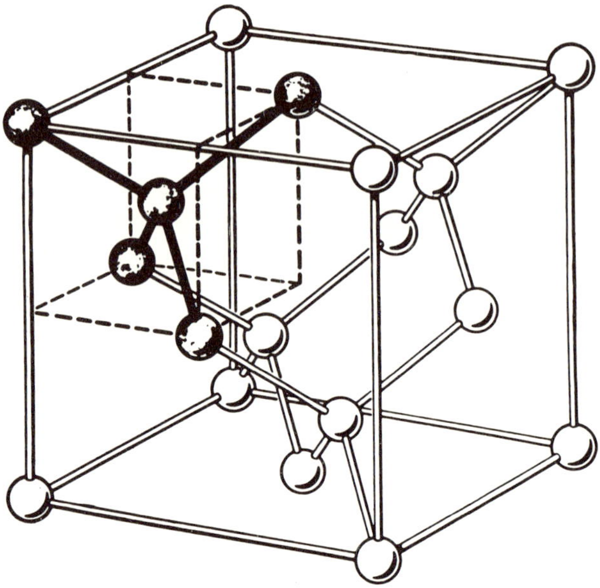

Figure 4. Crystal structure of diamond. (After W. Shockley, "Electrons and Holes in Semiconductors," 1950. D. Van Nostrand Company, Inc., Princeton, N.J.)

around a fifth so that all four balls are in completely equivalent positions with respect to each other.

This structure arises from the nature of the chemical forces that exist between atoms. The atoms in most substances are held together largely by electron-pair bonds resulting from the sharing of electrons between atoms. These electron pairs constitute the so-called covalent bonds.

A carbon atom, a silicon atom, and a germanium atom all have in common the fact that they have four valence electrons and form sets of four bonds with their four neighbors. In a perfect crystal, each covalent bond contains two electrons (Figure 5). Consequently, every electron is tightly bound and thus unable to enter into the process of electric conduction. This situation may be represented by the analogy of an automobile-storage garage, where one floor is completely filled with automobiles, so that no flow of traffic is possible. However, conductivity can be produced in crystals of this type in a number of ways, all of which involve destroying the perfection of the covalent bond structure.

The crystal in the perfect state may be compared to a perfect vacuum and is of no practical significance. A vacuum tube operates only because its vacuum is not quite perfect; electrons emitted by the hot filament make it potentially useful.

Various types of imperfections can be introduced in a semiconductor. For example, an electronic disturbance can be produced in germanium (Figure 6) when a photon delivers its energy to an electron, which is ejected from one of the bonds. This ejected electron constitutes a localized negative charge in the crystal, since before it arrived at its new

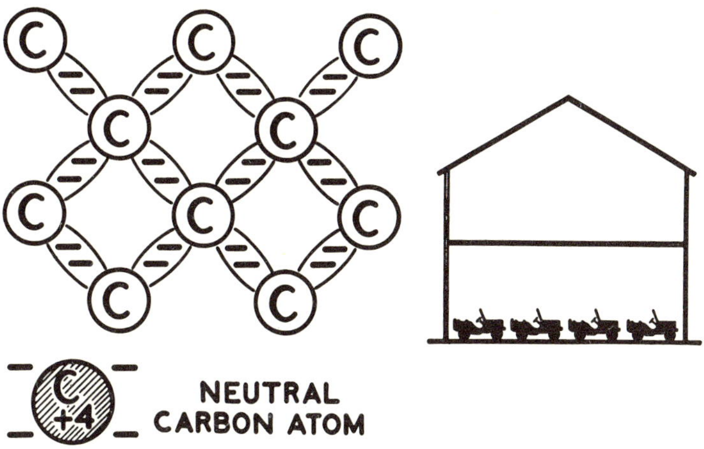

Figure 5. Electron-pair bonds in the diamond crystal. (Adapted from W. Shockley, "Electrons and Holes in Semiconductors," 1950. D. Van Nostrand Company, Inc., Princeton, N.J.)

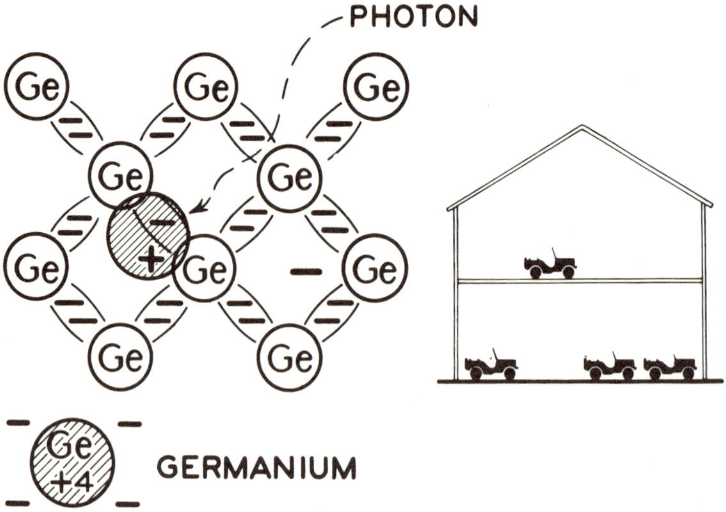

Figure 6. Intrinsic germanium: free electrons and holes produced by light. (Adapted from W. Shockley, "Electrons and Holes in Semiconductors," 1950. D. Van Nostrand Company, Inc., Princeton, N.J.)

destination in the crystal, the electron-pair bond structure was electrically neutral. Such an electron represents an excess over and above the number of electrons required to complete the bond structure in its neighborhood. It can move around much like an electron in a metal, since it has been brought into a set of higher energy levels, which in semiconductors are called the "conduction band." Its behavior is represented by the vehicle on the second floor, which is now free to move. This process of conduction by excess electrons is referred to simply as conduction by "electrons." In the automobile analogy, the vacant space now permits traffic to flow on the first floor. A similar conduction process takes place in the crystal through the motion of the hole left in the bond when the electron was ejected. This hole constitutes a net, localized, positive charge in the crystal, since before it was created that part of the crystal was electrically neutral. Experiments have shown unambiguously that holes move about as easily as electrons do, although this might not be intuitively obvious.

The conduction electrons and holes represent two types of imperfections, which can be introduced into the crystal by light. They can also be introduced chemically by substituting an arsenic atom, for example, for a germanium atom (Figure 7). The arsenic atom uses four of its five valence electrons to make four covalent bonds. Its fifth electron is free to wander about, giving the crystal a conductivity by negative carriers. This type of conduction is called *n*-type conduction, the crystal is called *n*-type material, and the arsenic impurity is called a donor-type impurity. Similarly, if an impurity with three valence electrons (like gallium) is substituted for a germanium atom in the crystal structure (Figure 8), it does not have sufficient electrons to form the four covalent bonds. Consequently, a hole is formed which is free to move about the crystal, giving it a conductivity by positive carriers. This type of conduction is called *p*-type conduction, the crystal is called *p*-type material, and the impurity is called an acceptor-type impurity.

These four types of imperfections represent the most important in transistor physics and transistor electronics (Figure 9). However, a fifth type of imperfection should also be mentioned, the so-called "deathnium," which acts as a recombination center for the minority carriers and consequently reduces the lifetime.

A useful analogy can be made between semiconductors and electrolytes (Figure 10). In a pure semiconductor, say germanium, containing no donors or acceptors, there is a certain density of electrons (n) and a certain density of holes (p). In water, due to electrolytic dissociation, there is a certain concentration of H^+ ions and OH^- ions. According to the mass action law, the product of the concentration of these two ions is constant at a given temperature, regardless of whether a base or an acid or a salt is present in the water.

The same is true for semiconductors. The electron concentration

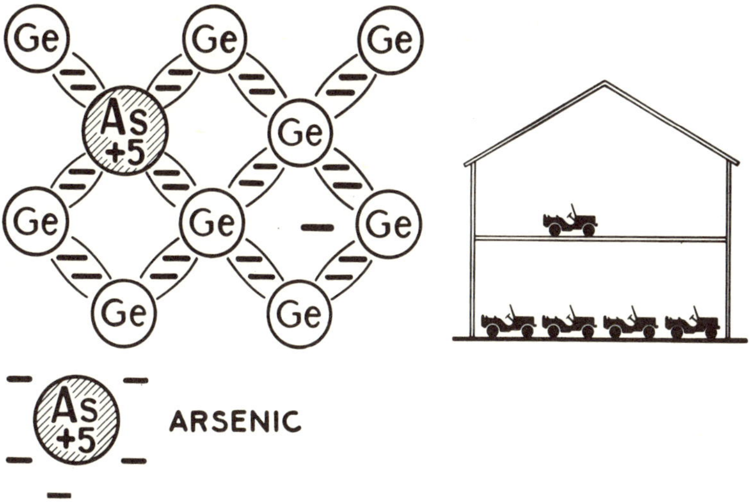

Figure 7. *n*-type germanium: electron conduction due to arsenic atoms serving as electron donors. (Adapted from W. Shockley, "Electrons and Holes in Semiconductors," 1950. D. Van Nostrand Company, Inc., Princeton, N.J.)

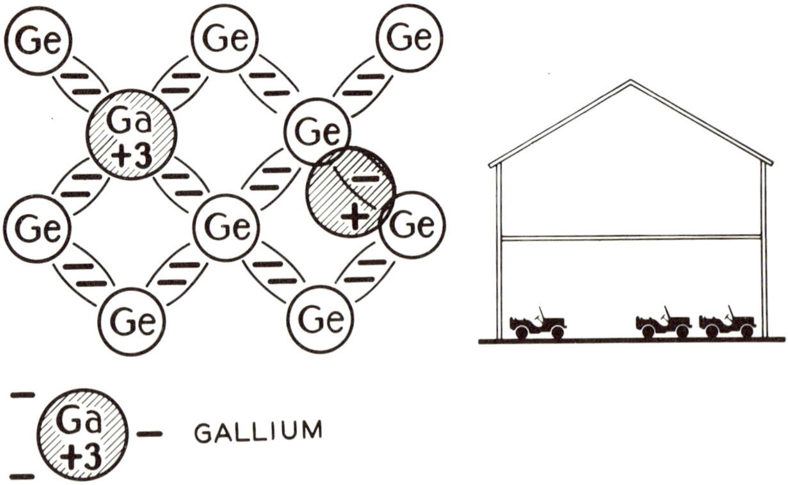

Figure 8. *p*-type germanium: hole conduction due to gallium atoms serving as electron acceptors. (Adapted from W. Shockley, "Electrons and Holes in Semiconductors," 1950. D. Van Nostrand Company, Inc., Princeton, N.J.)

IMPERFECTION	SYMBOL	RESULT
ELECTRON	−	n-TYPE CONDUCTIVITY
DONOR (element of valence 5)	⊕	
HOLE	+	p-TYPE CONDUCTIVITY
ACCEPTOR (element of valence 3)	⊖	
"DEATHNIUM" OR RECOMBINATION CENTER (Example: nickel in germanium)	□	REDUCED LIFETIME OF MINORITY CARRIER OR PHOTOCONDUCTIVITY

Figure 9. Types of imperfections in semiconductor physics.

SEMICONDUCTOR	ELECTROLYTE
INTRINSIC	WATER
$np = n_i^2$	$[OH^-][H^+] = K$
n – TYPE	BASE
$n = p + N_D$	$[OH^-] = [H^+] + [Na^+]$
p – TYPE	ACID
$n + N_A = p$	$[OH^-] + [Cl^-] = [H^+]$
COMPENSATED	SALT
$n + N_A = p + N_D$	$[OH^-] + [Cl^-] = [H^+] + [Na^+]$

Figure 10. Semiconductors and electrolytes.

times the hole concentration is constant at a given temperature and, in fact, is equal to the square of the electron concentration in the pure undisturbed crystal at thermal equilibrium, since in a pure sample the hole concentration is equal to the electron concentration n_i.

An *n*-type semiconductor is analogous to a base. In order to preserve neutrality in a base such as NaOH, the concentration of Na^+ ions plus the concentration of H^+ ions must equal the concentration of OH^- ions. The concentration of Na^+ is very much larger than the equilibrium concentration of H^+ in pure water. And since the product of the concentrations $[H^+]$ $[OH^-]$ is constant, $[H^+]$ must be much smaller than in pure water. Thus, in a base there are a negligible number of H^+ ions; that is, the hydrogen-ion concentration is suppressed. Similarly, in an *n*-type semiconductor, the concentration of holes is suppressed. Correspondingly, in a *p*-type semiconductor, the concentration of electrons is suppressed, just as the OH^- ion is in an acid.

There is a further important analogy to be drawn between semiconductors and electrolytes. An acid can be neutralized by a base, or vice versa. Similarly, if in germanium both donors and acceptors are introduced, one does not obtain the sum of the two conductivities. In fact, if introduced in equal amounts, the conductivity effects of the two types of impurities are canceled out. This process is known as "compensating."

3

This account of the evolution of the transistor principle and of the development of operating transistors must necessarily have a personal tinge. My interest in solid-state amplifiers goes back to the years before the Second World War. In 1938, Alan Holden and I attempted to make a solid-state amplifier using a piezoelectric crystal. Such a crystal produces a voltage when it is bent, and it is bent when a voltage is applied to it. The attempt to produce a solid-state amplifier by controlling the voltage applied to a device like a carbon microphone was not successful then, and neither of us has returned to it.

About 1940, on the basis of the rectifier theory of Schottky, I thought of a device closely related to what is now called the "field-effect transistor." After the war, I was able to prove that, in principle, it could be used as an amplifying valve controlling the flow of energy (Figure 11*a*). In this early field-effect transistor, the input is a condenser plate near the *n*-type semiconductor, the donors and electrons being represented by the symbols previously described. The battery is the source of energy trying to force electrons through the semiconductor; the ease with which electrons can be pushed through depends on the number

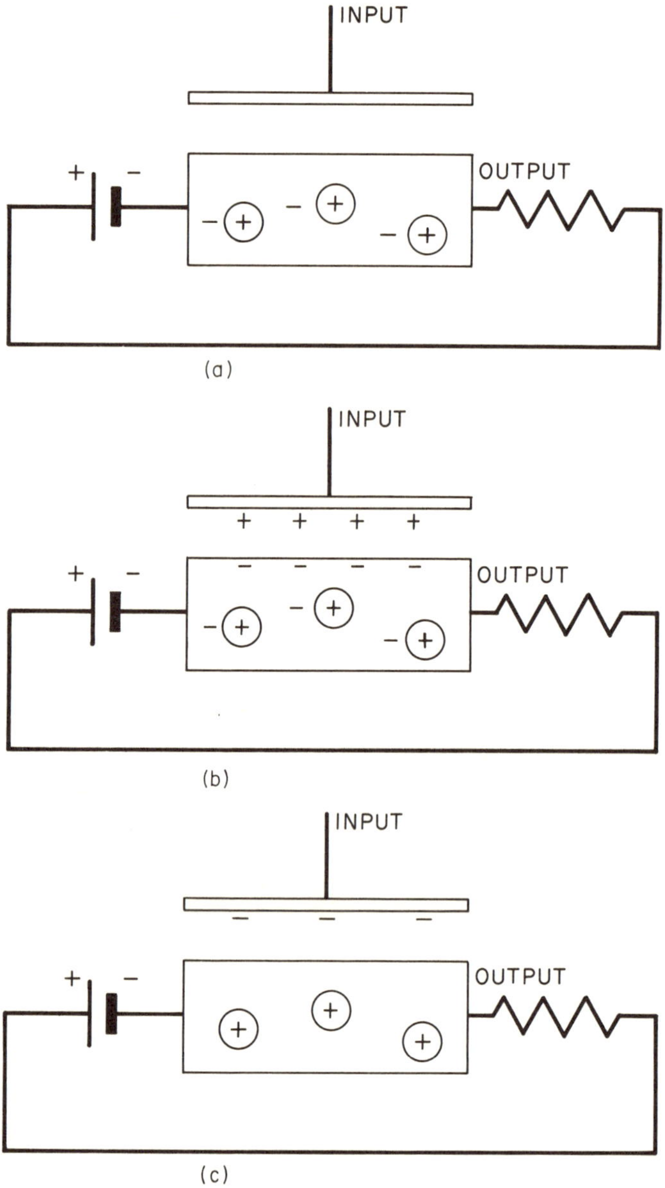

Figure 11. Schematic representation of the original field-effect transistor. The input condenser plate is shown (*a*) uncharged, (*b*) positively charged, (*c*) negatively charged.

of electrons in the semiconductor. The current flows through the output resistor, which may be a loudspeaker winding or any other appropriate load. If the input condenser plate is charged with a positive charge, additional negative charges will be induced in the semiconductor in the form of extra electrons (Figure 11*b*). Because there are more electrons in the semiconductor, more current can be forced through it by the electric field. In other words, the valve has been opened. If, on the other hand, the input condenser plate is charged with a negative charge (Figure 11*c*), positive charges will be induced in the semiconductor by repelling electrons, and perhaps the valve can be closed.

Three or four experimentalists successively were persuaded to try to make this device work or, at least, to observe an effect. But, by and large, all attempts met with negligible success.

At this point, John Bardeen suggested [1] that these extra induced electrons do not behave like the ordinary electrons in the bulk of the semiconductor. They are induced at the surface, which is unlike the bulk of the material because of the existence of "surface states," where these electrons can be trapped. He postulated that most of the induced electrons are, in fact, trapped in the surface states. Hence no measurable effect could be observed.

The surface states proposed by Bardeen were not supposed to behave like those we had theoretically predicted.[2] We thus put aside the work on an amplifier or an electronic valve and initiated a study on surface states. For about a year, many new experiments were devised and carried out, largely by Walter Brattain. These experiments, involving electrolytes, the change in contact potential with light, and other concepts, suggested that we could perhaps reach through these surface states and achieve control of the valve action in the semiconductor.

After a series of experimental and theoretical investigations, Bardeen and Brattain [3] announced in 1948 that they had obtained amplification with the point-contact, or type A, transistor. The point-contact transistor has two point contacts and an ohmic contact to a piece of *n*-type germanium (Figure 12). If one point contact is in the input circuit and the other point contact in the output circuit, the ohmic base contact being common to both circuits, amplification is achieved.

The development of the point-contact transistor illustrates an important principle which became evident in the course of our semiconductor research and which is probably relevant in many other branches of modern research.

We had made several approaches toward producing a workable field-effect device, but did not succeed. After three unsuccessful at-

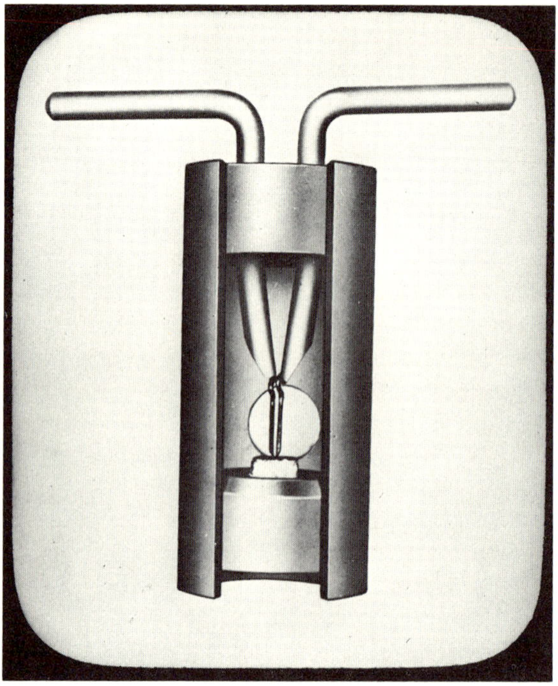

Figure 12. Cutaway view of a point-contact transistor. (Adapted from W. Shockley, "Electrons and Holes in Semiconductors," 1950. D. Van Nostrand Company, Inc., Princeton, N.J.)

tempts, we decided to concentrate on the scientific nature, rather than the practical aspects, of the problem. Bardeen's idea of the surface states then followed. In the course of these basic scientific studies, there were developed data and concepts which led Bardeen and Brattain to achieve the point-contact transistor. Thus respect for the scientific fundamentals of the practical problem and investigation of the reasons for the failure of the original field-effect device led to an experimental and theoretical program, which finally led not only to new scientific results, but also to a practical answer.

Another striking illustration of this principle is provided by some work done by Dr. Goetzberger,[4] a member of our staff in California. While investigating silicon *p-n* junctions (the compositional structure is described below), he found that these junctions permitted much more current to pass in the reverse direction than they should. The current-voltage characteristic of such a "soft" junction is shown in Figure 13. Not only is the current excessive (10^{-4} amp at about 10 volts), but also, as seen from the $-70°C$ curve, an important part of this current is not thermally generated. We tried various ways to reduce this excessive reverse current but soon found ourselves studying the phenomenon causing it. In other words, we turned from the practical to the scientific nature of the problem. After investigating several hypotheses, we finally arrived at the thought that the solution of the metal impurities in the silicon, at the elevated temperatures at which the junctions are made, becomes supersaturated as the silicon is cooled, and that the impurities precipitate as little metal particles. These particles in the space-charge layer might draw the excess current.

On the basis of this hypothesis, Goetzberger proceeded deliberately to perform experiments to increase further the excess current. He succeeded. If the hypothesis was correct, however, this excess leakage current should not be a surface leakage current, produced where the junction strikes the surface, but a leakage current produced within the bulk of the material.

In order to prove that it was not a surface current, metal was precipitated on a large-area *p-n* junction produced by diffusion. The *n*-type silicon was then removed everywhere, except in a 1-mm-diameter mesa (Figure 14). Three probes were placed on the mesa. Two hundred microamperes of reverse current was introduced on one probe; another

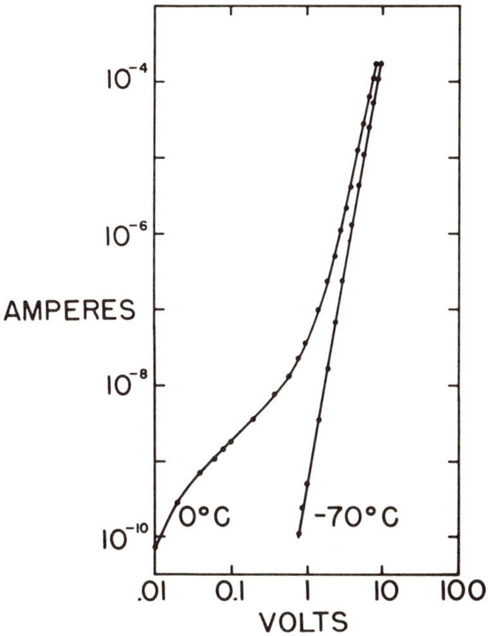

Figure 13. Current-voltage characteristic of a "soft" junction.

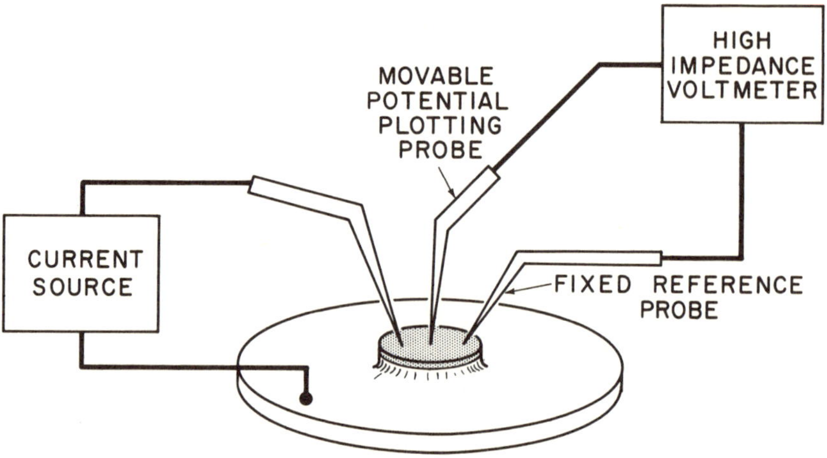

Figure 14. Schematic representation of experimental mapping of potential over the surface of a mesa diode.

probe was used as a fixed voltage reference; and the third probe was used to obtain a map of potential profiles over the surface of the mesa. This surface should be more or less an equipotential, since the junction is parallel to the surface.

Figure 15 shows the potential variations over the surface of the mesa. The current lines are perpendicular to the equipotential lines drawn. Thus all the current leaving the source point runs across the mesa surface, converging at one point, which is not where the junction meets the surface. This shows then that the excess reverse current is not a surface effect; it is a volume effect and matches the metal precipitate model.

Now, understanding the underlying mechanism, Goetzberger took steps to eliminate the metal precipitates. He placed a glassy oxide layer on the silicon, thus eliminating the metal precipitation at the time of formation. This process is called "gettering." This work has been very useful in the semiconductor industry as a means of avoiding and eliminating the contaminations that can be introduced during ordinary processing.

4

There was probably no more important scientific development in the semiconductor field in the early days following the announcement of the transistor than the development of high-quality single crystals of germanium at Bell Telephone Laboratories. A strong impetus for this work came from the transistor development department headed by J. A. Morton, who gave support to a program, previously urged by G. K. Teal, of making such crystals, using the Czochralski process. The first publications of success with this process were published by Teal and Little.[5] A crystal grown by this technique is shown in Figures 16 and 17.

If this crystal is grown from a melt containing arsenic, it becomes *n*-type. It is very difficult to remove the arsenic from the melt, but if part way through the growth one adds enough gallium, which is an acceptor, to more than compensate for the arsenic impurity, one obtains a *p-n* junction (Figure 18). This is a compositional structure; the making of these structures, in which the concentration and nature of chemical impurities vary from place to place, is most important in the technology of transistors and other semiconductor devices. They may be produced in a number of ways. It should be pointed out, however, that even after doping, these semiconductor crystals are chemically quite pure. Impurity concentrations of one part per million represent relatively heavy doping.

In a *p-n* junction (Figure 19), the *p*-type region is populated by holes and the *n*-type region by electrons. Consistent with the mass ac-

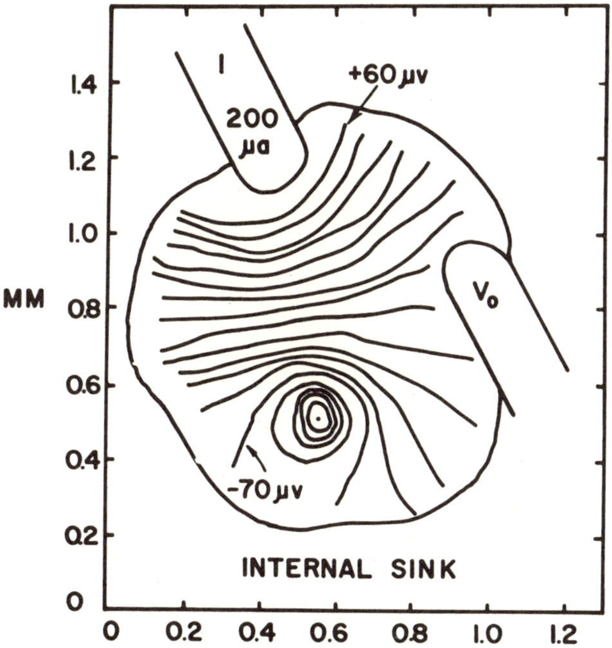

Figure 15. Potential distribution plot of the surface of a mesa diode containing precipitated metal particles in the space-charge layer.

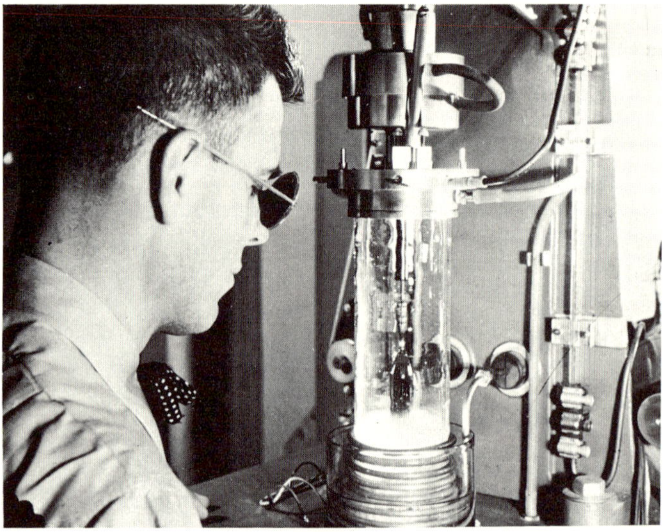

Figure 16. A single crystal of germanium being pulled from the melt, which is heated by induction. (Courtesy of Bell Telephone Laboratories.)

Figure 17. A typical germanium crystal, approximately $2\frac{1}{2}$ in. long, prepared by the Czochralski technique. (Courtesy of Bell Telephone Laboratories.)

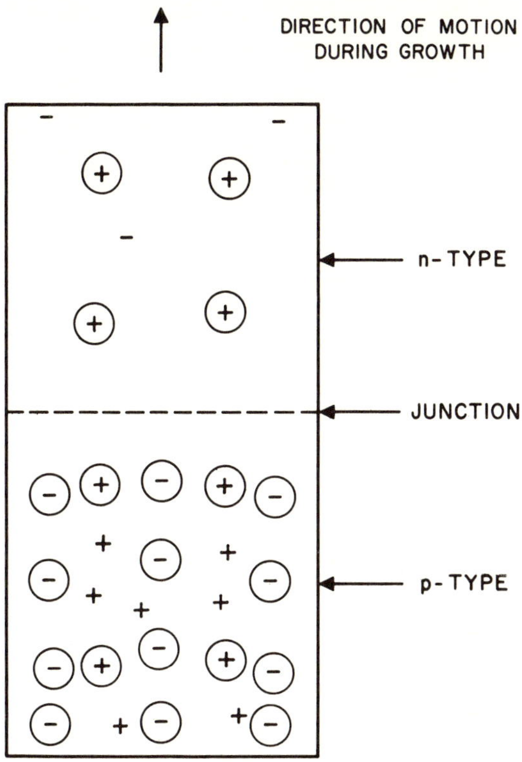

Figure 18. A *p-n* junction grown from the melt.

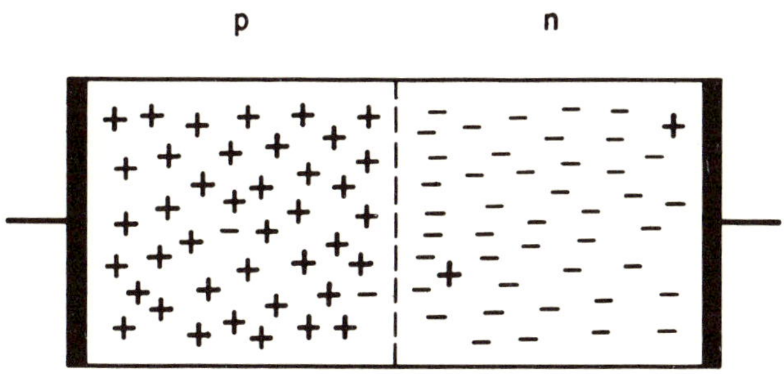

Figure 19. A *p-n* junction. For clarity, the donor and acceptor symbols have been omitted.

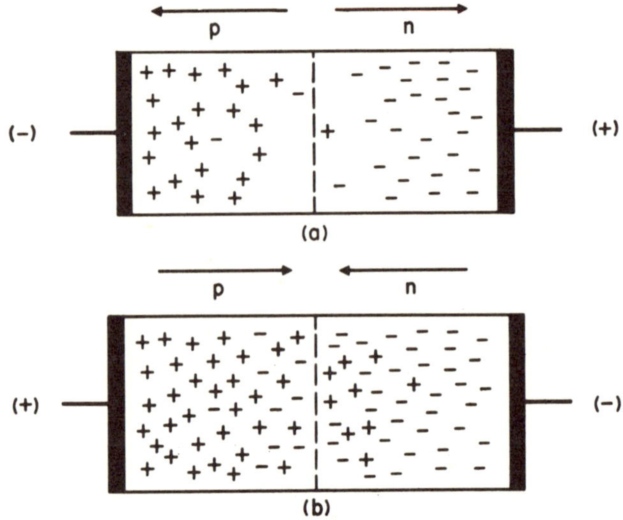

Figure 20. A *p-n* junction. (*a*) Under reverse bias; (*b*) under forward bias.

tion law, there are some holes in the *n*-type region and some electrons in the *p*-type region. This *p-n* junction is a photoelectric generator, the solar cell, and also a rectifier. The latter aspect of the *p-n* junction will be discussed briefly. In such a junction (Figure 20*a*), the *p*-type region has a negative bias with respect to one *n*-type region. The holes are attracted toward the negative side, and the electrons toward the positive side. Both types of carriers are pulled away from the junction, leaving ionized donors and acceptors which set up an electric field, thus producing a voltage drop across the junction. There is very little electric current flowing across the junction because the carriers are being pulled away from each other. This current is due to the occasional generation of some electrons in the *p*-type region near the junction and of some holes in the *n*-type region.

If the polarity of the voltage is changed (Figure 20*b*), the electrons and holes are pushed toward each other, recombine, and the device carries a large current. Figure 21 shows the voltage-current characteristics of such a *p-n* junction. The reverse-bias curve corresponds to the voltage polarity of Figure 20*a*, and the forward-bias curve to that of Figure 20*b*.

The theoretical curves of Figure 21 depict the well-known equation

$$I = I_0(e^{qV/kT} - 1) \tag{1}$$

where I and V are the current density and voltage, respectively, both being positive in the forward bias and negative in the reverse bias;

q is the electron charge; k is the Boltzmann constant; and T is the absolute temperature.

The ratio $kT/q \equiv V_0$ is the thermal energy and is equal to 25 mv at room temperature. This figure is a very important number; it determines the extent to which the forward current can increase as the forward bias is increased. At large forward currents, the Boltzmann factor $e^{qV/kT} = e^{V/V_0}$, or $e^{V/0.025} = e^{39V} = 10^{17V} = 10^{V/0.059}$. It should be noted that if the voltage is increased by 25 mv, the current can increase at most by a factor of $e = 2.73$. Alternatively, for every increase in the voltage of 59 mv, the current can increase at most by a decade.

It is evident that a p-n junction can be used as a rectifier. As shown in Figure 21, appreciably more current can flow in forward bias than in reverse bias. Thus, if an a-c voltage is applied across the p-n junction, appreciable d-c current flows in the forward direction.

5

The shortcomings of the field-effect-transistor concept can be overcome by replacing the condenser plate by a p-n junction.[6] In doing so, the limitations imposed by the surface states are minimized.

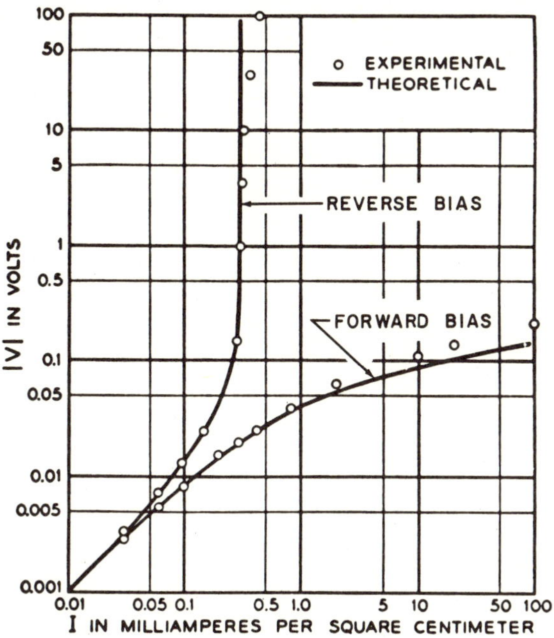

Figure 21. Voltage-current characteristics of a p-n junction. Experimental data are compared with theoretical curves from Equation (1).

Figure 22 represents an example of a unipolar field-effect transistor. It is a three-terminal device, consisting principally of a layer of an *n*-type material sandwiched between two layers of *p*-type material, the impurity doping being much stronger in the *p*-type material, which is designated as p^+. The working current is carried by electrons in a channel in the *n*-type layer between the two heavily doped *n*-type terminals designated as n^+. This compositional structure can be made in various ways, one of which involves the diffusion of acceptors into an *n*-type specimen.

The ease with which the battery can force the working current

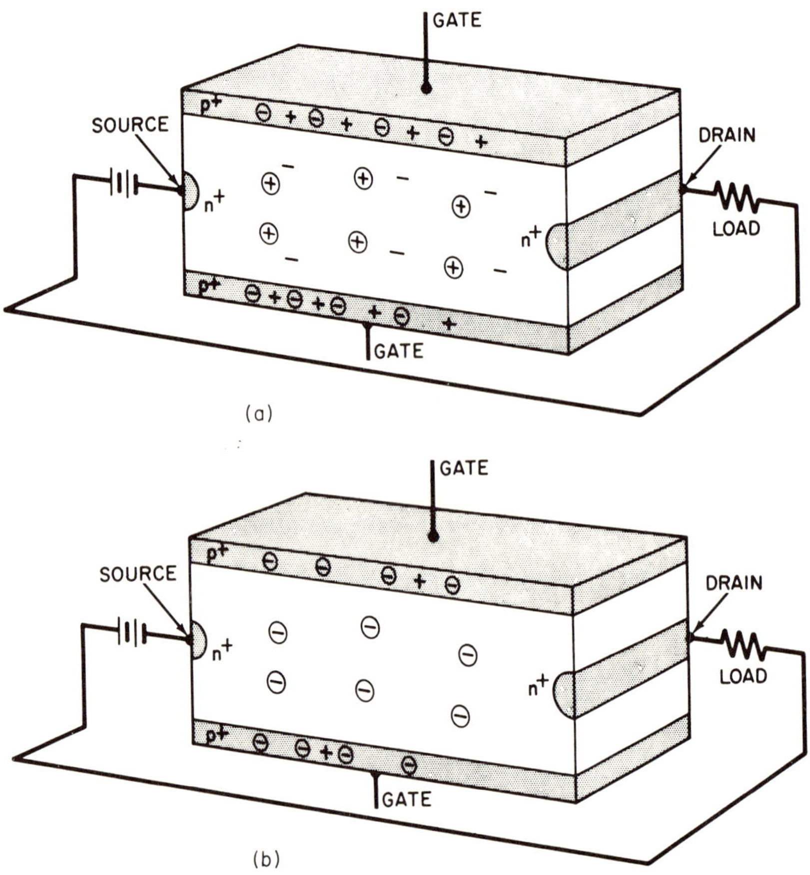

Figure 22. Unipolar field-effect transistor. The working current is carried between source and drain. The input signal is applied between the gate and the source. (*a*) No input signal; (*b*) negative input signal.

through the *n*-type semiconductor depends on the number of electrons in this semiconductor. In Figure 22*a*, for illustrative purposes, six donors and six electrons are shown in this *n*-type region, and four acceptors and four holes in each *p*-type region, called a gate. To modulate the working current, a negative charge is placed on the gates by removing holes from these *p*-type regions. When the holes are removed, the remaining unbalanced acceptors, which are negative, repel electrons, equal to the number of holes, which are removed from the *n*-type region (the crystal must be maintained electrically neutral). Thus, if six of the holes are removed from the gates of Figure 22*a*, all the electrons are removed in the *n*-type region as shown in Figure 22*b*, and this region becomes an insulator. Hence the surface states have not prevented valve action from taking place. Control has been exerted by having one form of electron flow, namely, the flow of holes in the gate region, control another form of electron flow, namely, flow of electrons through the channel.

In various ways, this process, in which one form of electron flow in the semiconductor controls and influences another flow of electrons, plays an essential role in practically all the modes of operation of transistors.

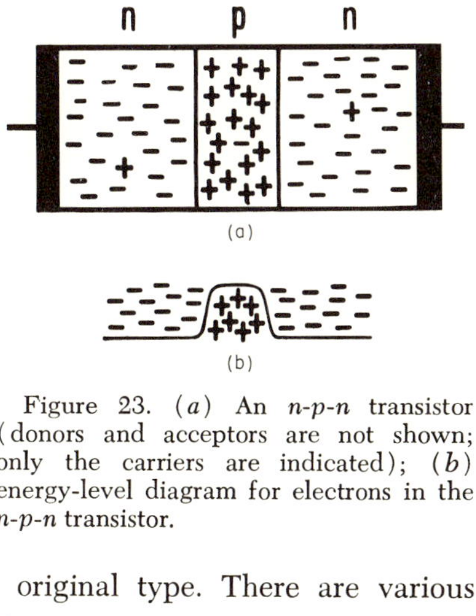

Figure 23. (*a*) An *n-p-n* transistor (donors and acceptors are not shown; only the carriers are indicated); (*b*) energy-level diagram for electrons in the *n-p-n* transistor.

The junction transistor is a compositional structure having three regions, *n*, *p*, and *n* (Figure 23*a*). The original *n-p-n* structure was prepared by first doping a melt with one impurity, then compensating it with the opposite type of impurity, and finally overcompensating with the original type. There are various ways of making this type of structure.

As shown in the energy-level diagram (Figure 23*b*), an electron finds the *n*-type regions, which are full of donors, energetically attractive; the holes (positively charged), on the other hand, find the *p*-type region energetically more attractive. Actually, the electrons must overcome a potential barrier in going from the *n*- to the *p*-type regions, and similarly, the holes must overcome a potential barrier in going from the *p*- to the *n*-type regions.

If voltage is applied in the *n-p-n* structure (Figure 24), the potential

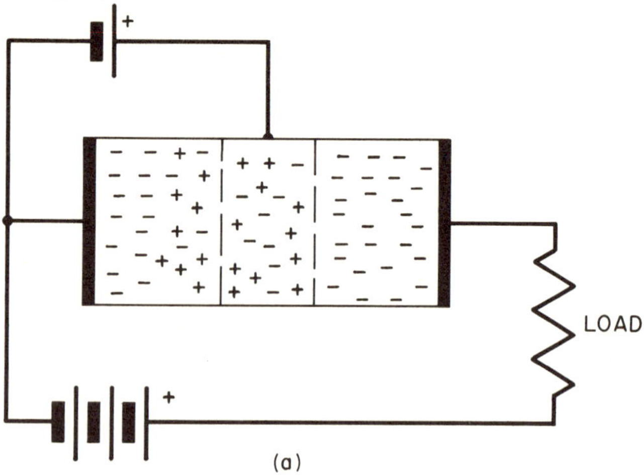

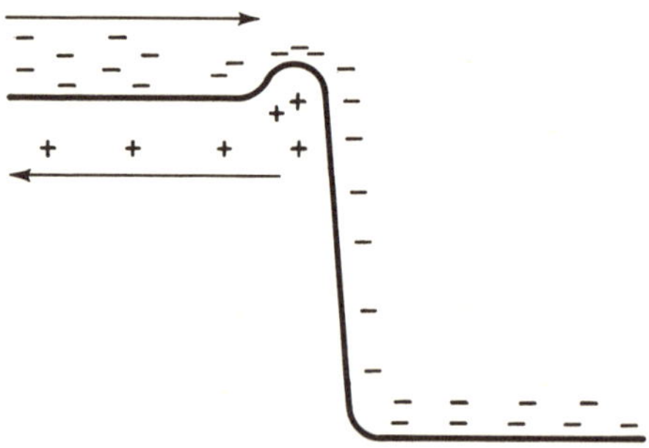

Figure 24. (*a*) An *n-p-n* transistor with voltages applied; (*b*) energy-level diagram for electrons in this transistor.

hill between the *p*-type region (base) and the right-hand-side *n*-type region (collector) is increased. An electron that reaches the edge of this hill slides down into the collector.

Normally, there are very few electrons in the base regions. If, however, the voltage is altered between the left-hand-side *n*-type region (emitter) and the base, the number of electrons crossing this *n-p* junction can be controlled. This voltage increases the number of holes in the base and helps electrons to be injected into the base. If the base is thin enough, and if there are not too many deathnium, or recombination, centers to cause recombinations of the electrons with holes, then a small number of holes added in the base can produce a large additional flow of electrons through the base.

Thus again, one kind of electron flow, namely, the hole flow, controls another, the electron flow. In this case, unlike the field-effect transistor, these two related currents are actually in the same space. They are separate, however, since they represent different mechanisms of flow of electricity.

The junction transistor is by far the most important transistor structure now made. It was first described in print from the point of view of theory in an article published in 1949.[7] The first publications describing outstanding achievement of its potential were published in 1951.[8]

A third type of transistor is the so-called filamentary transistor (Figure 25). This transistor[9] consists of a bar of, say, *n*-type silicon with ohmic contacts at both ends. The resistance between these contacts is modulated by injecting holes in the *n* region from a *p*-*n* junction. These holes introduce a positive charge in the *n* region and attract additional electrons. For every extra hole in the *n* region, there is an extra electron, and the resistance to electron flow between the two end contacts is reduced. Thus the process results in amplification.

6

In the field-effect transistor, the junction transistor, and filamentary transistor, one electron flow controls another. There are, however, other amplifying principles in the field of solid-state electronics, one of which is amplification in volume.

The cryosar[10] (low-temperature *s*witching by *a*valanche and *r*ecom-

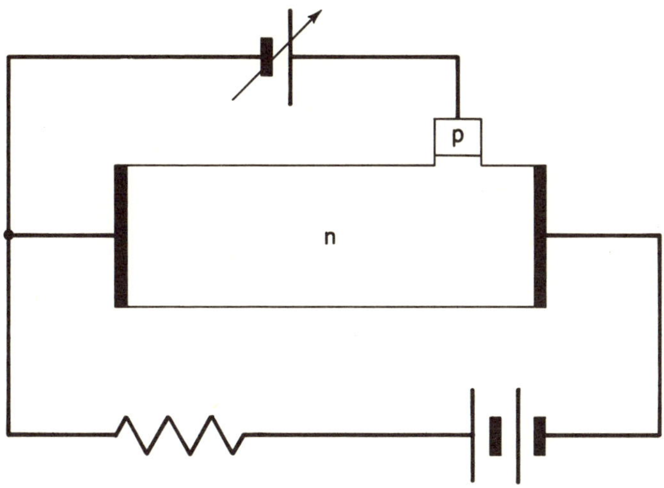

Figure 25. Filamentary transistor. The working current is carried between the end contacts. The input signal is applied to the *p*-*n* junction.

bination), which was pioneered at the Lincoln Laboratory of M.I.T., is based on the properties of semiconductors at low temperatures. As the temperature of a semiconductor is lowered, the number of free electrons is reduced because they become attached to donors (Figure 26). Thus, at very low temperatures, the semiconductor becomes less a semiconductor and more an insulator. If an electric field is applied to the semiconductor, the very few electrons which have remained free can acquire sufficient energy to free, by impact, some of the bound electrons and thus initiate an avalanche process very similar to the gas-discharge ionization process. Such a process can provide switching action and power gain.

An entirely different principle of amplification in volume has led to maser amplifiers [11] (*m*icrowave *a*mplification by *s*timulated *e*mission of *r*adiation), in which the intensity of electromagnetic radiation is amplified by making use of the appropriate electron levels in the bulk of a crystal.

One can determine whether there is energy dissipation or energy generation within a volume by integrating the divergence of Poynting's vector over the volume or by integrating the vector itself over a surface enclosing the volume (Figure 27). Power is generated where the divergence of Poynting's vector is positive. In the amplifier shown in Figure 28, the equations indicate that if Poynting's vector is integrated correctly over a surface enclosing the amplifier, it is found that the difference between the power input (including the d-c source power) and the power output is the power dissipated in heat. The a-c and d-c components of Poynting's vector can be separated; the a-c power can be said to be generated where the a-c component of Poyn-

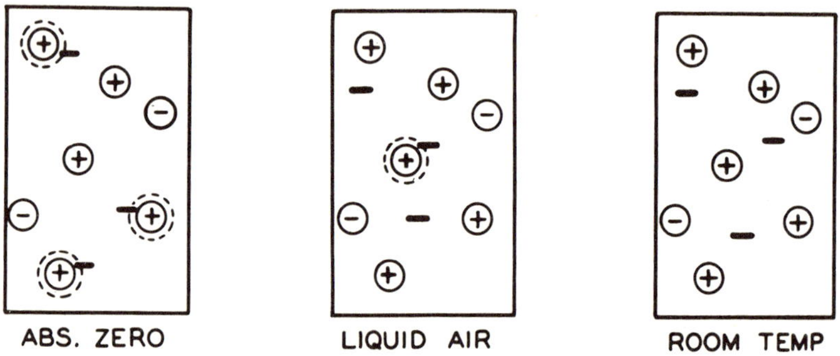

Figure 26. Effect of low temperatures on the number of free electrons in *n*-type silicon. (Adapted from W. Shockley, "Electrons and Holes in Semiconductors," 1950. D. Van Nostrand Company, Inc., Princeton, N.J.)

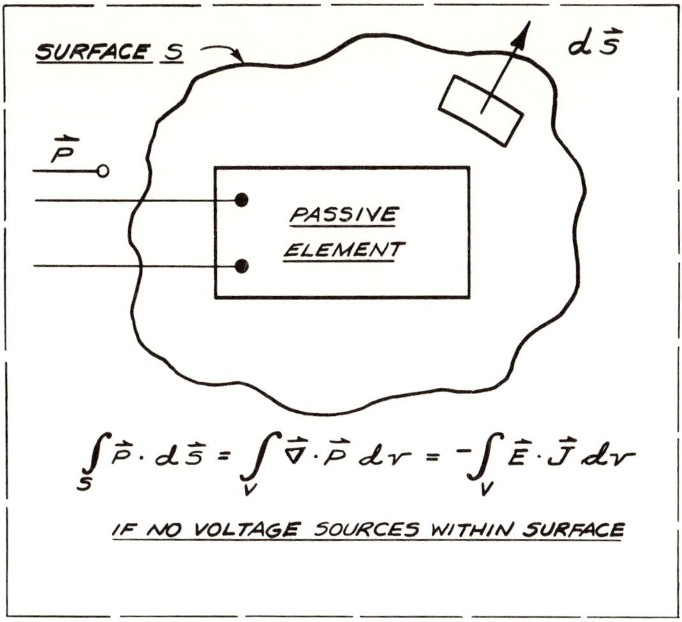

Figure 27. Relationship between Poynting's vector and energy dissipation.

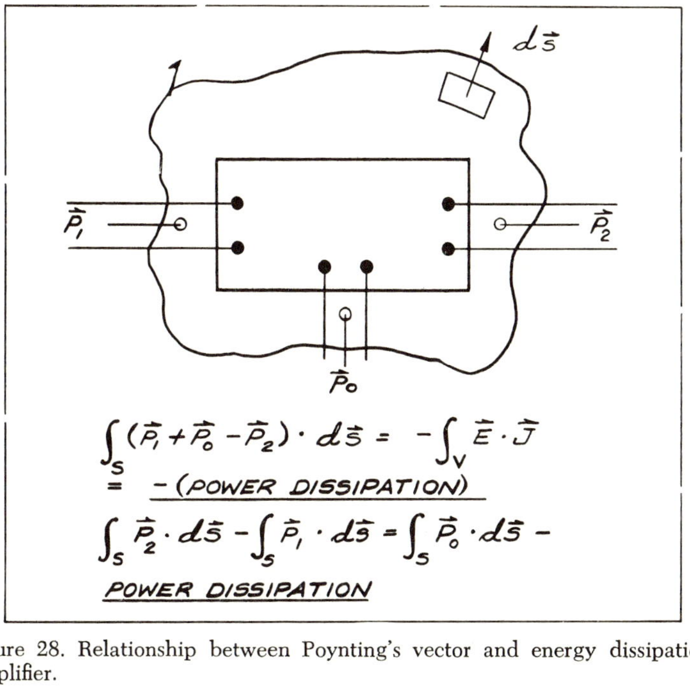

Figure 28. Relationship between Poynting's vector and energy dissipation for an amplifier.

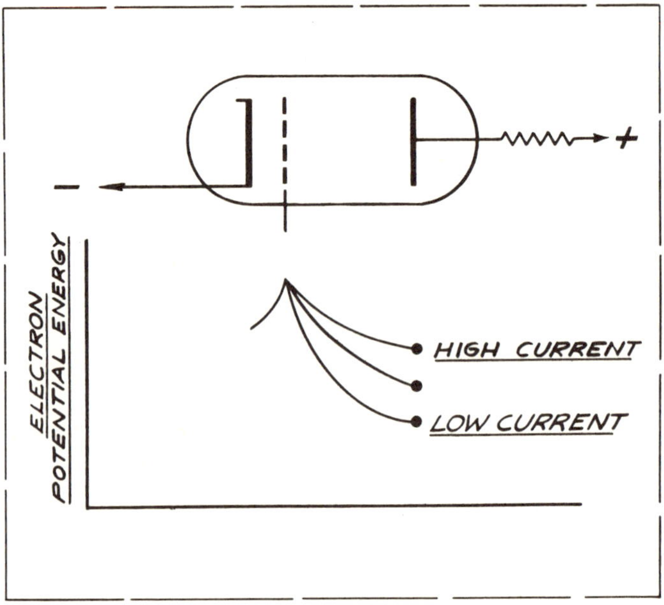

Figure 29. Electron potential energy as a function of distance in a vacuum tube for different values of current.

ting's vector has a positive divergence. Figure 29 shows where this power is generated in a vacuum tube. In any element of volume between the grid and the plate, the product of the a-c components of current and voltage is negative because, when the current is large, the voltage in the direction to make the current flow is small. Now since

$$\int_s \rho \cdot ds = \int_v \nabla \rho \, dv = -\int_v \mathbf{E} \cdot \mathbf{J} \, dv \qquad (2)$$

if the product of $\mathbf{E} \cdot \mathbf{J}$ is negative, Poynting's vector is positive, and consequently power is generated. Thus a-c power is generated in the vacuum tube in the region between the grid and the plate.

In the collector space-charge region of a junction transistor, when the current is large, the voltage in the direction to make the current flow is also small (Figure 30). Thus it is in this region, where the product of the a-c components of $\mathbf{E} \cdot \mathbf{J}$ is negative, that the a-c component of Poynting's vector is positive and that the a-c power is generated.

It may be appropriate to speculate at this point about the future of transistor electronics. Those who have worked intensively in the field share

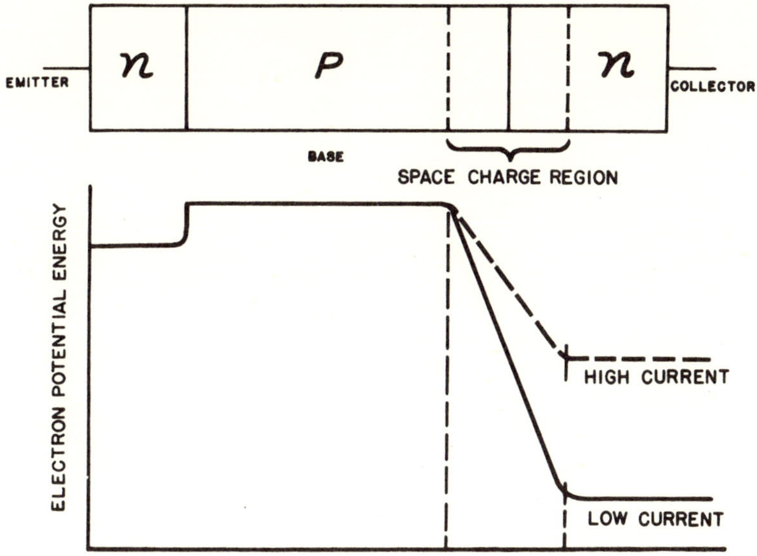

Figure 30. Electron potential energy as a function of distance in a junction transistor for different values of current.

the author's feeling of great optimism regarding the ultimate potentialities. It appears to most of the workers that an area has been opened up comparable to the entire area of vacuum and gas discharge electronics. Already several transistor structures have been developed and many others have been explored to the extent of demonstrating their ultimate practicality, and still other ideas have been produced which have yet to be subjected to adequate experimental tests. It seems likely that many inventions unforeseen at present will be made based on the principles of carrier injection, the field effect, the Suhl effect, and the properties of rectifying junctions. It is quite probable that other new physical principles will also be utilized to practical ends as the art develops.

This prophecy was made in my book [12] in 1950. More than ten years later, it still seems a reasonable forecast of things to come.

REFERENCES

1. J. Bardeen, Surface States and Rectification at a Metal to Semiconductor Contact, *Phys. Rev.*, vol. 71, pp. 717–727, 1947.
2. W. Shockley, On the Surface States Associated with a Periodic Potential, *Phys. Rev.*, vol. 56, pp. 317–323, 1939.
3. J. Bardeen and W. H. Brattain, The Transistor, a Semiconductor Triode, *Phys. Rev.*, vol. 74, pp. 230–231, 1948; Physical Principles Involved in Transistor Action, *Phys. Rev.*, vol. 75, pp. 203–231, 1949.

4. A. Goetzberger and W. Shockley, Metal Precipitates in Silicon *p-n* Junctions, *J. Appl. Phys.*, vol. 31, pp. 1821–1824, 1960.
5. G. K. Teal and J. B. Little, Growth of Germanium Single Crystals, *Phys. Rev.*, vol. 78, p. 647, 1950.
6. W. Shockley, Unipolar "Field Effect" Transistor, *Proc. IRE*, vol. 40, p. 1365, 1952.
   G. C. Dacey and I. M. Ross, The Field Effect Transistor, *Bell System Tech. J.*, vol. 34, p. 1149, 1955.
7. W. Shockley, The Theory of *p-n* Junctions in Semiconductors and *p-n* Junction Transistors, *Bell System Tech. J.*, vol. 28, pp. 435–489, 1949.
8. W. Shockley, M. Sparks, and G. K. Teal, *p-n* Junction Transistors, *Phys. Rev.*, vol. 83, pp. 151–162, 1951.
   R. L. Wallace, Jr., and W. J. Pietenpol, Some Circuit Properties and Applications of *n-p-n* Transistors, *Bell System Tech. J.*, vol. 30, pp. 530–563, 1951.
9. W. Shockley, G. L. Pearson, and J. R. Haynes, Hole Injection in Germanium—Quantitative Studies and Filamentary Transistors, *Bell System Tech. J.*, vol. 28, pp. 344–366, 1949.
10. A. L. McWhorter and R. H. Rediker, The Cryosar—A New Low-temperature Computer Component, *Proc. IRE*, vol. 47, pp. 1207–1213, 1959.
11. J. P. Gordon, H. J. Zeiger, and C. H. Townes, The Maser—New Type of Microwave Amplifier, Frequency Standard, and Spectrometer, *Phys. Rev.*, vol. 99, pp. 1264–1274, 1955.
12. W. Shockley, "Electrons and Holes in Semiconductors," Van Nostrand, Princeton, N.J., 1950.

Charles H. Townes holds the fundamental patent on the maser. He was Professor of Physics at Columbia University and Vice President and Director of Research of the Institute for Defense Analyses before assuming his present position as Provost, Massachusetts Institute of Technology.

CHAPTER 8

MASERS

1

Electronics is a young field, and perhaps subject to even more rapid change than most fields of science or technology. One visible and likely stimulus for change in the character of electronics during the next decade is the maser, which has been taking form and gradually emerging from research laboratories during the last decade. The maser is a device based on a new type of amplification, with properties and potentialities almost qualitatively different from what has previously been available. It plays a key role in a new field sometimes called "quantum electronics," which is concerned with the production and control of electromagnetic waves by their interaction with electrons bound in atoms or molecules, instead of those flowing through a vacuum as in a vacuum tube, or through a semiconductor as in a transistor. This takes us into a world that was once of interest only to the fundamental sciences—the structure of atoms and molecules, spectroscopy, and quantum mechanics.

The phenomenon of amplification itself may be characterized as a controlled release of stored energy. In the case of the maser, energy is stored in the excitation of individual atoms and molecules. It is released by subjecting the molecules to electromagnetic waves of just the right frequency which stimulate the molecules to emit their excess energy. The released energy is given up as electromagnetic radiation to the stimulating wave, which is thus amplified. Hence the name maser, which was originally an acronym for *m*icrowave *a*mplification by *s*timulated *e*mission of *r*adiation. The technique is by no means limited, however, to microwaves and has now been extended over such a wide range of frequencies that the name should perhaps be interpreted as *m*olecular *a*mplification by *s*timulated *e*mission of *r*adiation, since molecules are the important amplifying element. It has also been suggested that the word "maser" has a cabalistic meaning of particular significance to laboratory directors and to government agencies, i.e., *m*eans for *a*cquiring *s*upport for *e*xpensive *r*esearch.

Considered as an invention, the maser epitomizes the great change that has recently come over the character of technological frontiers. It was worked out and predicted almost entirely on the basis of theoretical ideas of a rather complex and abstract nature. This is not an invention or development which could grow out of a basement workshop, or solely from the Edisonian approach of intuitive trial and error; it is rather a creature of our present scientific age which has come rather completely from modern physical theory.

The remarkable properties and versatility of this new device offer exciting possibilities to a wide variety of fields. As amplifiers of radio waves, masers are 100 to 1,000 times more sensitive than those which were previously available. We thus suddenly have amplifiers which no longer contribute significant noise and which increase markedly the possibility of long-distance communications and radar. The use of maser amplifiers in the West Ford terminals has enabled Lincoln Laboratory to establish a 50,000 bit/sec transcontinental communication link via the moon. Frederick Kappel, president of American Telephone and Telegraph, has said, "Without the maser we do not see how we could even hope for high-quality space communications." [1]

Large radio telescopes already probe somewhat farther into the universe than the most powerful optical telescopes. Because of greater sensitivity, maser amplifiers on these telescopes can increase even their range by a factor of almost 10, allowing us to approach still closer to the edges of the universe. Figure 1 shows an early maser used for radio astronomy,[2] placed at the focus of a 50-ft dish so that it can amplify the very weak radio waves picked up from outer space. This particular maser is neither small nor cheap. But the potential gain in sensitivity would allow the 50-ft telescope as great a range as a 500-ft telescope using an ordinary amplifier. Maser amplifiers on radio telescopes have already produced improved measurements of the temperatures of planets, and the first measurements of the amount of hydrogen in a number of galaxies.

A quite different facet of masers is seen in the atomic clock. Not very long ago, the best clocks accumulated errors at the rate of about 1 sec every 10 years. Maser oscillators operate with such constant

Figure 1. National Research Laboratory 50-ft radio telescope with maser installed at the prime focus. (Courtesy of National Research Laboratory.)

frequency that they can be used to control clocks which accumulate errors at a rate of 1 sec in 10,000 years. There are now hopes for maser clocks of still greater accuracy.

Another important aspect of masers, which is still rather different from the last two, is that they provide amplifiers and oscillators at frequencies which are enormously higher than those previously attainable with electronic tubes. They allow us to expand electronic-like techniques from the normal radio region into the infrared and optical regions of the electromagnetic spectrum—and probably beyond. Their making available this expanded frequency range to refined and flexible techniques has been compared in importance with opening up of the microwave region, on which so much of present-day radar and communications is based. Masers operating in the optical and infrared region give us a source of light which is many orders of magnitude more intense than that of the sun, many orders of magnitude more directive than any previous searchlight or radio beam, and with a frequency purity as good as the best electronic oscillators. Optical and infrared masers are beginning to be used for such diverse fields as radar, communications, microscopy, welding, and eye surgery. Research effort in this field exploded from almost zero three years ago to a current effort involving several hundred different companies and laboratories.

2

With these indications of the significance of masers in mind, let us now take a more leisurely and historical approach to their development. It is perhaps premature to view masers from a good historical vantage point since they are still quite new; the most active research in this field is probably still yet to come, and we are just in the midst of exploring their potentialities. However, we have at least an opportunity to discuss our subject with fresh memory of its early stages, to examine an interesting development in mid-course, and to attempt a projection toward the future. What I have to say about this development must necessarily be based on the experiences and evolving ideas of an individual in the midst of his own hobby. But perhaps it will nevertheless give some helpful perspective and useful illustrations of typical processes involved in scientific and technological progress.

All too frequently, we look back in time at some invention or discovery and contract its development into a sudden revelation to one particular man, and at one particular moment. This makes invention and scientific discovery appear spectacular and mystical, when in fact it is carried out by ordinary human beings, in small and stumbling steps. What really is marvelous is that scientific knowledge is so fruitfully accumulative. One idea or technique is added to another, making still a third one easier to understand or achieve. Soon a structure has developed which is, in fact, impressive and spectacular, even though individual steps viewed in detail are simple and ordinary.† This is clearly not meant to belittle the scientist, engineer, or inventor, but rather to emphasize the extent to which science consists of the accumulation of small, uncertain steps. If this seems belittling, it is perhaps most appropriate that I illustrate by my own field. It is also important to be aware of the extent to which science and engineering are social phenomena, evolutionary processes involving the cultural medium, the interaction between many individuals, and the hand-in-hand development of both fundamental and applied science.

† Cf. Aristotle, "Metaphysics," A.10, 993[a] 30: "The search for truth is in one way hard and in another easy. For it is evident that no one can master it fully nor miss it wholly. But each one of us adds a little to our knowledge of nature and from all the facts assembled there arises a certain grandeur."

In order to discuss the maser and how it amplifies electromagnetic waves, it will perhaps be helpful to review the character of these waves. As indicated in Figure 2, they include a number of familiar types of radiation (light, X rays, radio waves) and great variety. But despite this variety, they are all similar in the sense that they obey the same physical laws, differing only in wavelength or frequency. One of the least known regions lies between the microwave and infrared frequencies because the only sources of radiation in this no-man's-land of the electromagnetic spectrum are hot objects, which radiate very weakly at these relatively long wavelengths, thus allowing only primi-

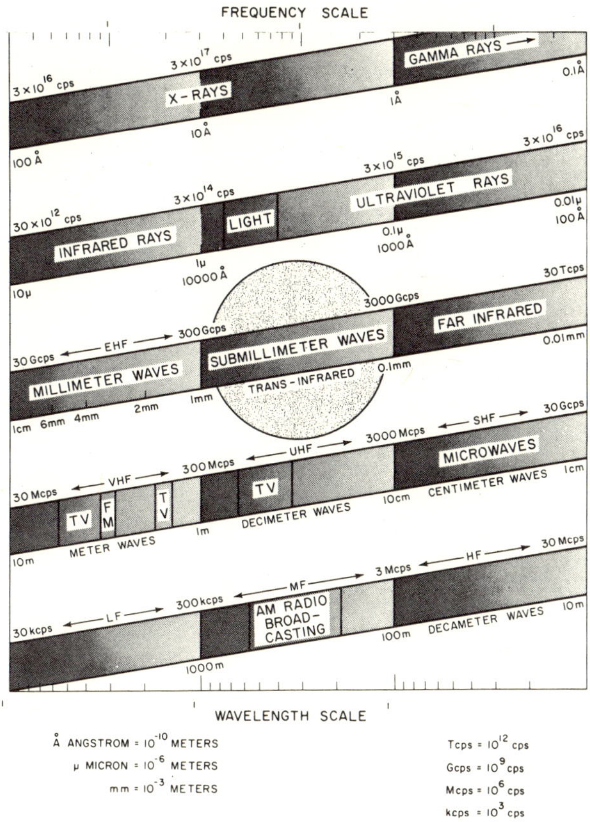

Figure 2. The spectrum of electromagnetic waves. The spectrum displayed on a logarithmic scale suffers an apparent bandwidth contraction as the frequency increases. The absolute bandwidth of any given area on the chart is as many orders of magnitude larger than the same area of a lower-frequency region as the difference in decades between the two regions. The region occupied by visible light, for example, has about the same bandwidth (approximately 4×10^{14} cps) as the entire spectrum lying below it in frequency.

tive study of the region. At higher frequencies, hot objects radiate more intensely, while at lower frequencies electronic oscillators or their harmonics are available as sources.

The maser amplifies electromagnetic waves by stimulating emission from atoms or molecules in ways well known to the physicist. The theory of atomic emission and absorption was developed as a result of intensive spectroscopic work, primarily in the optical range, over many decades. Niels Bohr first suggested that light or other forms of electromagnetic radiation is emitted when an atom makes a transition between two discrete energy levels. If the atom falls from an upper level to a lower level, it emits a quantum of electromagnetic energy (Figure 3*a*). If it absorbs a quantum of electromagnetic energy from a wave, it must make a transition from the lower level to the upper level (Figure 3*b*).

In 1917, Einstein published an important paper on radiation which gave a clear exposition of the relation between the various processes by which atoms absorb and radiate electromagnetic waves. He pointed out that there must be two separate ways by which an atom may emit a quantum of radiation. It may fall spontaneously from a higher to a lower energy level, producing electromagnetic waves where none was previously present. This spontaneous emission is responsible for essentially all the light we see from the sun or from incandescent lamps. If there is present, however, an electromagnetic wave of the appropriate frequency, it may stimulate an atom in the upper state to fall to the lower state, giving up its energy to the stimulating wave. The appropriate frequency, as Bohr pointed out, is such that the frequency multiplied by Planck's constant is equal to the difference in energy of the two levels. These ideas are basic to spectroscopy and to the emission or absorption by atoms of electromagnetic waves of all frequencies.

Rather recently physicists have extended spectroscopic research into the microwave and radio regions. Perhaps Cleeton and Williams [3] in

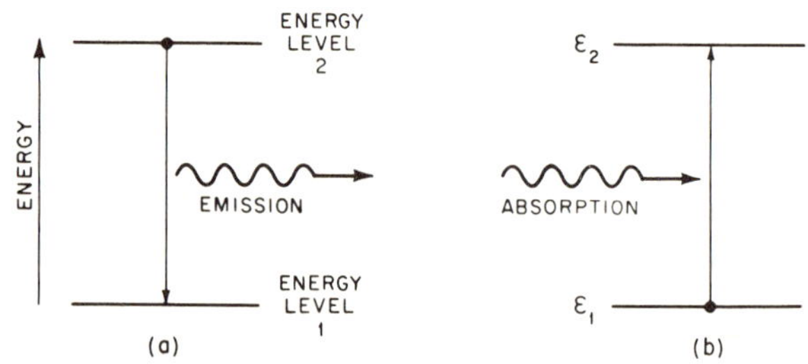

Figure 3. Energy levels of atomic systems. (*a*) Emission of radiation; (*b*) absorption of radiation.

1934 were the first to demonstrate a reasonably sharp spectroscopic resonance in response to waves in the radio region when they examined the absorption of microwaves in ammonia. I. I. Rabi's work on radio-frequency resonances in beams of atoms came also before World War II and led to much interesting research. But very extensive study of spectroscopy in the radio and microwave regions was to come only later as a result of intensive development of microwave technology during the war and the familiarity with electronics and microwaves which this gave many physicists.

A glimpse into the state of affairs by early 1945 is given by an internal memorandum [4] of a large industrial laboratory, written to convince the research director that microwave spectroscopy was of both practical and scientific interest.

> Microwave radio has now been extended to such short wavelengths that it has overlapped a region rich in molecular resonances, where quantum mechanical theory and spectroscopic techniques can provide aids to radio engineering. Resonant molecules may furnish a number of circuit elements of future systems using electromagnetic waves shorter than one centimeter. Many difficulties in manufacture of conventional circuit elements for very short wavelengths can be obviated by using molecules, resonant elements provided by nature in great variety and with the reproducibility inherent in molecular structure.
>
> So little work has been done in microwave spectroscopy, which we define broadly as the interaction of microwaves and matter, that the course of its future development and application is difficult to predict. Knowledge of the possibilities in this field comes largely from extrapolation of studies in the far infrared, from Rabi's work at wavelengths in the meter region, and from the few resonances of oxygen, water vapor, and ammonia which have received some attention in radar propagation investigation.

The memorandum expresses, with some foresight, high hopes for the general usefulness of molecular spectroscopy in technology. But the details were by no means clear. Devices such as the maser were certainly not imagined, as can be seen from a paragraph further along in this memorandum.

> Molecular gas may be heated or otherwise excited to emit discrete bands of radiation whose frequency width may be adjusted over wide limits. . . . Although this application probably warrants investigation, it does not appear immediately promising because the radiation produced would be incoherent and because of intensity limitations. In most cases the molecules would dissociate before very great excitation is obtained.

The latter statement was generally accepted by scientists of the time and was based on the argument that, although molecules may produce radio-frequency waves or other types of radiation in rather narrow bands, thermodynamic reasoning showed that the intensity of this radiation could never be greater than that produced by a black body (that is, by normal, hot material), which is small at radio frequencies. The argument was sound, in general, but there were loopholes, and the lack of a clear understanding of these loopholes created a stumbling block to progress toward the maser. But there did follow an active and fruitful development of microwave spectroscopy of gases, solids, and liquids.

Microwave techniques had been carried to a remarkably high level during World War II. Their subsequent application to spectroscopy interested many scientists and furnished a great deal of valuable fundamental information, which whetted interest in extending microwave techniques to still shorter wavelengths. In 1945, the shortest microwaves produced were about 5 mm in length. My own interest, as well as that of many other spectroscopists, was much attracted to the frequency range between microwaves and the infrared, because this no-man's-land could be predicted to be rich in spectra and because it represented a challenge. Applied scientists were also strongly interested because it was an unexploited frequency range, with characteristics somewhat similar to those of the very useful microwaves.

Early in 1950, the Office of Naval Research sponsored a committee of scientists and engineers to attempt to stimulate interest in the production of waves in the millimeter and submillimeter range and to help provide and screen ideas for this purpose. The Carbide and Carbon Chemical Corporation also became interested in very short waves for stimulating specific chemical reactions and supported research toward their generation at Columbia University.

After almost two seasons of service on the ONR committee, I felt that some progress had been made, but that good sources of very short waves still seemed far in the future. In the spring of 1951, I was in Washington to attend the next meeting of the committee, and found myself sitting on a bench in Franklin Park early one morning, admiring the azaleas, then at the height of their bloom, but also wondering whether there was a real key to the production of very short electromagnetic waves. Suddenly I realized what was needed.

Clearly, for the shortest waves, we must rely on atomic and molecular resonators, since the appropriate resonators are too small for man to construct. The thermodynamic argument that molecules could not radiate more than a black body was faulty because it implied equilibrium, a state easily obviated, for example, by molecular-beam techniques, with which some of my Columbia colleagues had been working for years. Perhaps I had in the back of my mind the recent radio-frequency experiments on nonequilibrium and stimulated emission carried out by Purcell and Pound [5] at Harvard. The feedback necessary for oscillation could be provided by a cavity containing the radiation. In a few minutes I had calculated, on the usual back of an envelope, the critical condition for oscillation in terms of the number of excited molecules which must be supplied and the maximum losses allowable in the cavity.

It was very exciting to find in principle a way of producing a very short-wave oscillator, even though it did not look easy or promising, and though I thought I had been stupid to have been diverted from this approach for so long by the thermodynamic argument. Actually, much of what was involved was well known to physicists.

If all the atoms or molecules in a given sample of matter are in the upper state of two energy levels, and if these molecules are stimulated by an electromagnetic wave of the right frequency, the wave will induce some downward transitions and gain energy from the molecules. Under equilibrium conditions, there are always more molecules in the lower than in the upper state. Since electromagnetic waves induce transitions upward as well as downward, for material in equilibrium there is a net flow of molecules to the upper state, which robs the wave of energy, producing absorption. But if there are more molecules in the upper state than in the lower, the stimulated transitions produce a net flow of molecules to the lower state, delivering energy to the wave. This much of the basic principles involved was understood by any physicist thoroughly familiar with radiation theory since the time of Einstein's paper. I can remember several informal discussions among physicists shortly after the war on proposed experiments to detect stimulated emission. One finds the possibility of "negative absorption" dealt with understandingly, though briefly, in several places beginning as early as the mid-thirties.[6]

Two specific things had previously been lacking: a mechanism for obtaining net amplification, that is, an arrangement which would actually produce amplification and oscillation, and strong motivation for attempting what seemed to be a very difficult experiment.

But also, for most new ventures, the time must be ripe, and this had apparently not then been the case for masers. It was the group of physicists engaged in radio and microwave spectroscopy after the war, immersed in a combination of spectroscopy and electronics, who seem to have had the background and habit of mind required; the principal advances have come almost exclusively from this group. The three independent serious suggestions for use of stimulated emission for amplification came from three teams, each working on the microwave spectroscopy of gases—our own group at Columbia University, A. Prokhorov and N. Basov at the Lebedev Institute in Moscow, and J. Weber at the University of Maryland. The two groups that tackled the problem of making circuit losses less than the gain supplied by molecules, and hence of obtaining net gain, ended up with remarkably similar schemes, using a beam of ammonia molecules fired into a low-loss resonant cavity. In view of the wide variety of masers now known to work (some much more easily than the original system), this can hardly be a random coincidence. Here surely one detects a pervasive atmosphere of the times. The success of our own device was also much dependent on a method for producing very intense molecular beams which had just been worked out by Paul and his associates in Germany and which I had learned about through his visit to New York.

In the original ammonia-beam maser (Figure 4), a beam of ammonia molecules is emitted through a small hole into a vacuum. An arrangement of electric fields pushes and pulls on the molecules in such a way that those in the ground state are pulled out of the beam and those in the excited state focused through a hole in a resonant microwave cavity, where they can interact with the microwaves and give them their energy.

It should be remembered that, although this new type of amplifier was an interesting idea, few regarded it at the time as a "hot" topic for

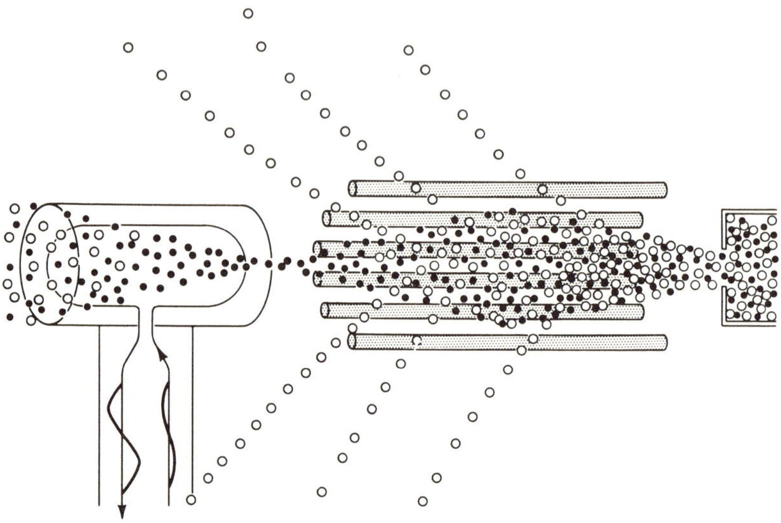

Figure 4. Ammonia-beam maser.

either physics or engineering, or gave it much priority. As a topic, its "temperature" has been gradually rising ever since. Whereas new ideas on masers are now sometimes overrushed to the editor of the *Physical Review,* things were different a decade ago, as may be judged from the long delay between active discussion of masers and their appearance in standard technical journals (see Reference 6 for details of such dates). As other documentation on how uncertain our steps toward progress can be, I might mention a visit from two of my most distinguished colleagues, who urged me to stop this silly work on a molecular oscillator, which seemed very unlikely to function, because it was wasting government money. The project was being generously supported under a joint services contract administered by the Signal Corps, and I had by that time spent about two years and $50,000 on this questionable device. Their objections were a help, because the natural, and perhaps calculated, result was that I pushed on the experiment harder than ever.

In late 1953, after somewhat more than two years of work, the maser oscillated, to the triumph and credit of James P. Gordon, then a graduate student, who had risked taking on this apparently very difficult experiment. H. J. Zeiger, who had also helped to get the experiment under way, had by that time moved to Lincoln Laboratory.

Even at this point, the new device was not received with unmitigated enthusiasm. Another of my good friends and colleagues, after hearing a seminar on the operation of the new oscillator, advised me that he thought some of my other work was really more worthwhile. What, after all, he asked, could I do with this oscillator, now that it did work? Although I had some ready replies, I'm sure they weren't particularly convincing.

Into new research areas, our view is indeed through a glass darkly. I believe that, with a little wisdom, research will yield useful results almost as surely as summer follows winter, but their precise nature is no more foreseeable than the details of next summer's weather. We had been searching for sources of radiation in the no-man's-land between microwaves and infrared. But we had so far turned up an oscillator of remarkable stability (and hence an atomic clock)—and the realization of what an excellent noise-free amplifier the maser might become. Now there are optical and infrared masers, but with all the maser's present successes, we still have none in this no-man's-land.

On the other hand, in 1951 one could already see, in a vague way, important applications for this amplifier and the possibility of a useful technology. A lecture in 1951, after some consideration of the beautiful scientific results obtained from microwave spectroscopy, makes the following comments: [7]

> We have been discussing the many types of fundamental information that have been obtained as a result of the technological development of radar. The historical development in this case is just the opposite of what most pure scientists like to regard as typical. They would argue that in most cases pure science develops principles and ideas, which are then applied to technology by other scientists and engineers and incorporated in equipment and devices. But in the case of microwave spectroscopy, the reverse has occurred. The equipment and devices were developed first, and pure science owes a considerable debt to technology. There is, however, every reason to believe that microwave spectroscopy will in time be able to repay this debt. It seems likely that work in this field will lead to the use of groups of molecules to generate, detect, attenuate, or filter microwaves. These groups of molecules can in many ways do at high frequencies what ordinary circuit elements, such as resistors, condensers, and inductances, do at lower frequencies.

Some of the broad outlines of the future were visible, but let me disclaim any great clairvoyance by also pointing out that, a year after the maser was working, I took a sabbatical leave with the plan of examining myself and physics and of deciding whether or not it would be wise to change to some other more useful type of research. I could have continued interesting enough work on the maser, but the next most useful steps were by no means clear.

Perhaps by good luck, when I settled in Paris I encountered Arnold Honig and Jean Combrisson, who were working with a paramagnetic impurity in germanium of very striking properties, which I had not previously realized could be achieved. An easy calculation showed that it could produce a maser amplifier, and I found myself again, late in 1955, in the happy state which most researchers know of losing sleep as the result of an exciting idea.

Whereas the ammonia-beam-type maser could amplify only over a very narrow range of frequencies and was practically untunable—properties which made it a very good constant-frequency oscillator—paramagnetic materials responded over a wide frequency band and were tunable, just as is needed for a good amplifier. The particular paramagnetic substance we were studying involved an electron spin which could be oriented in two directions in a magnetic field, giving it the possibility of two different energies. The separation between these two levels, and hence the frequency which might be amplified, could be changed simply by varying the magnetic field strength. Fortunately, friends at the Bell Telephone Laboratories were skilled at making the germanium necessary and were glad to supply this material when I explained our aim of making a maser amplifier in Paris. In the short time available there, we managed only to make preliminary tests and to show clearly that the material could be made to amplify.[8] Amplification was later achieved with this material by Gordon, Feher, and others at the Bell Telephone Laboratories.[9]

But such ideas were ripening in the scientific community, and very soon a somewhat different and improved approach paid off handsomely. I believe that it was in the spring of 1956 that Strandberg at M.I.T. realized independently that paramagnetic materials would make good maser amplifiers and was actively considering the use of two energy levels in such materials. A talk he gave on this possibility aroused the interest of Bloembergen at Harvard, who had been working for some time on paramagnetic phenomena. Bloembergen had an even better idea: the use of paramagnetic material with three energy levels in a suitable way would provide a more convenient and usable amplifier operating on the maser principle.[10]

The three-level solid-state maser can be understood from Figure 5. In equilibrium, the upper of the three levels is occupied by very few molecules, the next level is occupied by a somewhat larger number, and the lowest level by a still larger number (Figure 5*a*). If an electromagnetic wave of frequency corresponding to the energy difference between the uppermost and lowest levels is present, molecules on the lower level will make transitions to the upper level, and vice versa, until their populations are nearly equal. This decreases the population of the lower level (Figure 5*b*). If the decrease is sufficient, the number of molecules in the lowest level may become less than that in the second level, and maser amplification is possible by stimulated emission (Figure 5*c*). This system is useful partly because it can amplify con-

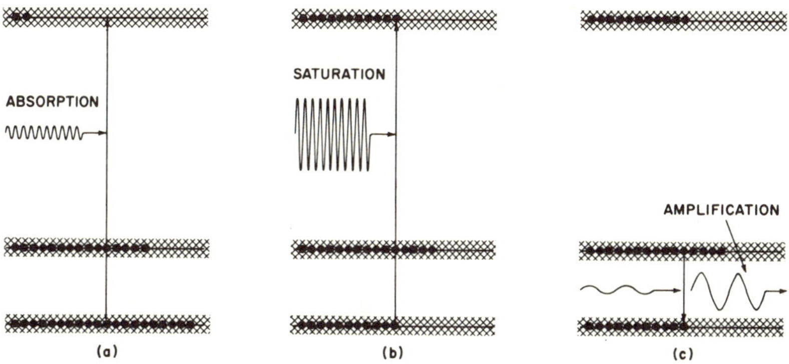

Figure 5. Three-level mode of maser operation. As suggested by Bloembergen.

tinuously; i.e., by sending in continuously the electromagnetic waves which force transitions from the lowest to the uppermost state, a steady-state condition in the flow of molecules can be achieved in which the central state is more densely populated than the ground state.

It is again evident that such ideas were in the air; one finds that they occurred, more or less independently, to microwave spectroscopists in several places.

Especially as a result of Bloembergen's suggested scheme, there followed a burst of activity to develop and utilize this new amplifier which theory indicated would be almost noise-free. By the fall of 1956, a number of laboratories were actively pursuing solid-state masers, and Scovil, Feher, and Seidel at the Bell Telephone Laboratories had produced the first solid-state maser.[11] Lincoln Laboratory was one of the leaders in this development; McWhorter and Meyer there were the first to measure certain of the new amplifier's characteristics; and the first application of a maser to a high-power radar was made by Kingston on Lincoln Laboratory's Millstone Hill system.

It should be noted that this burst of activity and successful use of solid-state masers would probably not have occurred if other scientific and technological developments had not produced the right setting. There was the considerable background of knowledge and understanding of paramagnetic resonances resulting from fundamental studies in microwave spectroscopy. There was also the increasing use of liquid helium brought about to a considerable extent by the Collins liquefier; a less convenient source of low temperature would have exercised considerable restraint on the use of solid-state masers. The availability of synthetic ruby, suggested as a favorable paramagnetic substance by Kikuchi at the University of Michigan, was likewise important. The development of radio astronomy and the possibility of long-range communciations by means of satellites provided much of the impetus for examination and utilization of these very low noise amplifiers. The development of ferrite materials and the invention of the circulator were important ingredients in exploitation of the enormous practical gain in sensitivity afforded by maser amplifiers.

For a while those interested in masers were happily inventing new types of solid-state amplifiers, designing and engineering older types, and getting them installed on antennas for some of their initial utilization. The region between microwave frequencies and infrared frequencies was still undeveloped, and one could only hope for some good ideas to emerge for masers in this region as the result of intense activity at longer wavelengths. By 1957, however, I was impatient with waiting for good ideas and sat down to think at some length about what might be done to make masers which would operate at these short wavelengths. Progress seemed to be hindered largely by two difficulties. One was the problem of building cavities to contain the electromagnetic waves at these short wavelengths which would not also have a large number of possible modes of oscillation and which would hence be confusing and difficult to control. The other was that spectra and techniques in the region were not very well known; since no available set of molecular resonances was more than barely marginal for maser operation, the problem of struggling also with difficult techniques seemed discouraging.

The solution to this muddle was perhaps to ignore the first difficulty and jump directly into the still shorter wavelengths of the optical region where spectra and techniques were already very well known. Calculations showed that masers could be made to operate in the optical region with known spectroscopic transitions and reflecting cavities as resonators. Such masers would be very interesting in themselves. Furthermore, together with microwave masers, they could provide the basis for an assault on the no-man's-land from either larger or shorter wavelengths. But it was in discussions with Schawlow of the Bell Telephone Laboratories that the best ideas came to be developed.[12] He had also been actively considering optical masers, and suggested that plane parallel mirrors would be helpful. The more they were examined, the more advantages appeared for a suitable arrangement of parallel mirrors.

Consider two plane parallel mirrors between which there is a collection of atoms in excited states. A single quantum of light emitted from one atom can stimulate emission from others. If the stimulated emission is in a direction such that it is reflected back and forth between the mirrors, and if the total amount of light emitted is greater than the loss in the wave due to reflection from the mirrors, then the wave gains in intensity during repeated reflections. Ultimately, its intensity is such that it saps the energy of all atoms fed into the space and cannot grow further. This is the regime of steady oscillation.

Light can be reflected straight back and forth between the two mirrors without leaking out because of its short wavelength. Hence this system provides a resonator which takes advantage of the short wavelength rather than being too handicapped by it. Analysis showed that a proper choice of mirror dimensions and molecular properties could solve the problem of multiple modes, which had seemed so troublesome. Waves traveling in directions other than straight back and forth represent most of the multiple modes. But they are eventually reflected out of the system, and hence are not amplified. Suitable choice of dimensions of the two mirrors and of the properties of the atoms allows one to ensure that the system oscillates in only one mode and gives only a single frequency. Such an oscillator is coherent and produces as pure a frequency as the best electronic oscillators, even though it is singing at optical frequencies near 10^{15} cps.

As with maser amplification, and as with the use of three levels and paramagnetic resonance, invention of the optical maser now seems in retrospect remarkably simple. The broad ideas are, in fact, very simple and represent only a small addition to what was previously understood. The details become rather more complicated, but not difficult to work out, once the appropriate train of thought is begun.

Analysis of these ideas and ways of obtaining excited atoms seemed to make it clear that optical and infrared masers would before long be made to work, and that they would have extremely interesting properties. Successful use of the spectral transitions of certain excited gaseous atoms, for which much was known, could be rather completely predicted by calculation; but the less well-known transitions in solids seemed perhaps even more promising. Excitement began to rise again, and very soon there were a large number of additional suggestions for supplying excited molecules and for designs of optical and infrared masers. However, any time there is overconfidence about how clearly recognizable the most useful research projects are, I have several counterbalancing anecdotes to relate about the reception this idea received at the hands of some intelligent physicists and patent lawyers.

Maiman [13] at the Hughes Research and Development Laboratory was the first to achieve experimentally maser action at optical frequencies. His method involved intense illumination of a piece of pink ruby with plane parallel faces. A group [14] at the Bell Telephone Laboratories quickly showed still more clearly than did Maiman's initial experiment that such a system (Figure 6) gives, in fact, the characteristics expected of maser oscillation. This maser operated at a rather high power level and in pulses only, because such high illumi-

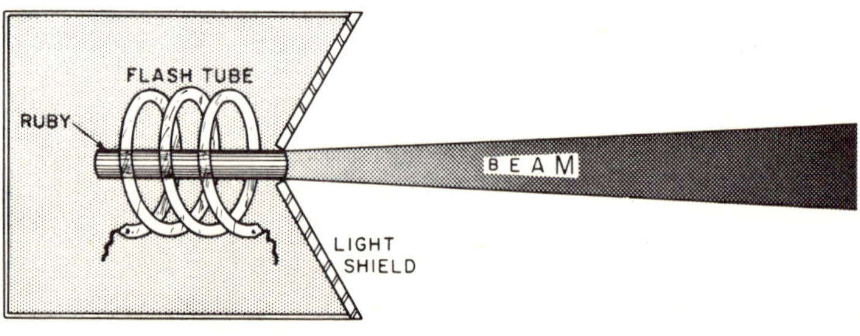

Figure 6. Ruby optical maser of the type achieved by Maiman.

nation intensity could not practically be maintained for more than a short time. The red ruby color is produced by a rather small number of impurity atoms of chromium in the hard crystal of aluminum oxide with energy levels indicated in Figure 7. Maiman realized that it was possible, with sufficient intensity of exciting light, to deplete markedly the lowest energy levels of the chromium atoms. When absorbed, blue light raises these impurity atoms from the ground state to a broad spectrum of excited states. The excited atoms then drop down to an intermediate level and accumulate there in such numbers that their population exceeds the population of the depleted ground state. The ruby is then ripe for maser action. A wide variety of crystals have now been made to operate by somewhat similar schemes.

Javan, then at the Bell Telephone Laboratories but now at M.I.T., suggested[15] the use of a gaseous discharge in which helium atoms in metastable states transfer energy to neon atoms (Figure 8). The discharge excites helium atoms, allowing accumulation of metastable helium. On collision with a neon atom, a metastable helium atom transfers energy, bringing it to an upper excited state; this produces populations in the upper states larger than those in intermediate states. Again maser action can occur. Javan, Bennett, and Herriott of the Bell Laboratories made such a system operate[16] (Figure 9). Its particular charm is that it oscillates continuously, can be rather precisely controlled, and gives performance close to the best that is theoretically expected of a maser.

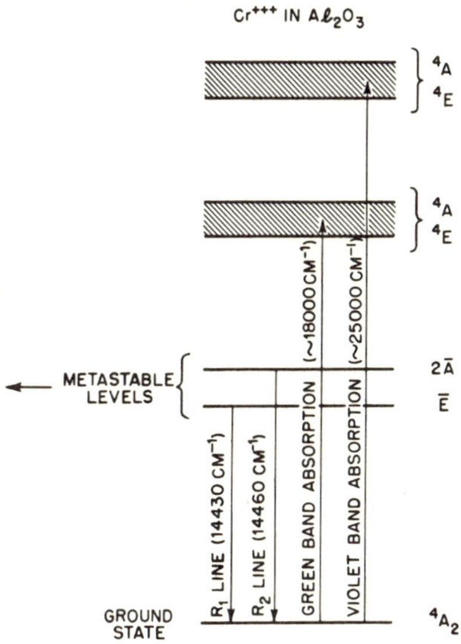

Figure 7. Energy levels of ruby.

The optical maser is often called a laser, for *l*ight *a*mplification by *s*timulated *e*mission of *r*adiation. In order to avoid proliferation into such terms as iraser for infrared, uraser for ultraviolet, and raser for radio-frequency masers (which have all been suggested), it seems desirable to retain maser as the general term, with the frequency range specified, as in infrared or optical maser. But laser is conveniently short, and not inappropriate for masers operating in the optical region.

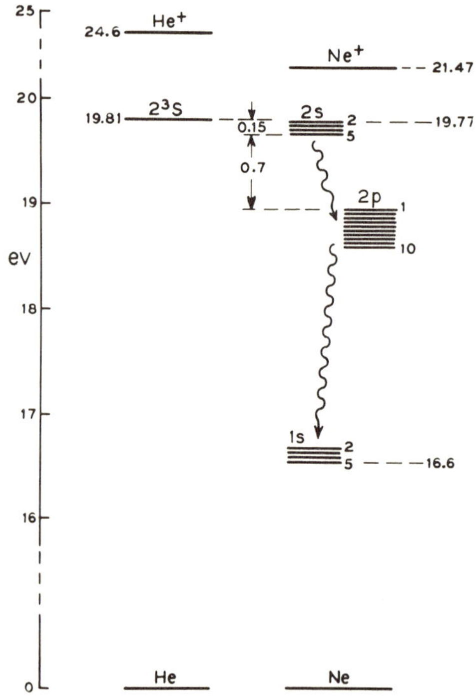

Figure 8. Energy states in the helium-neon mixture.

By now optical and infrared masers are the subject of widespread and intensive research, and many types have been built. Why they were this long in being discovered is a little difficult to say. Consider the components of Javan's maser, for example. It consists only of a gaseous discharge in helium and neon and two parallel mirrors. The basic quantum mechanics and spectroscopy required to understand its operation were well known thirty years ago. Why wasn't it invented at that time, in the great age of optical spectroscopy? Somehow the trend of thought had to be connected with electronics where coherent oscillators are commonplace, and to go through microwave spectroscopy, then microwave masers, and finally back to the older fields of optics, gaseous discharges, and atomic spectroscopy.

3

We have seen some of the many steps in the development of masers, mostly in themselves rather simple, and the scientific and technological climate which made them possible. Let us now take time for a broader view of where this has brought us. Man's pathway seems much more impressive when thus viewed broadly, either backward or forward in time, rather than in detail. Where now do we stand in the development of masers, and what further are the theoretical potentialities?

One can show theoretically that a maser amplifier should produce so little noise that it can detect an electromagnetic wave containing only one quantum of energy. It has been the custom to express the noise properties of an amplifier in terms of a "noise temperature," that is, the temperature of an object emitting heat radiation into an amplifier at a level which doubles the amplifier noise output. In these terms, good amplifiers in the microwave region normally have noise temperatures in the range of 1000 to 2000°K. In terms of quanta of energy, this means that the amplifier can detect a minimum of about 10,000 quanta of microwave radiation. By comparison, the theoretical noise temperature of a maser amplifier in the microwave region is somewhat below 1°K, which is so low that it no longer enters into any engineering considerations. A maser amplifier has been shown experimentally to have a noise temperature less than 2°K. This is simply the best performance which can at present be measured and indicates that no amplifier noise has been detected at all. It corresponds to detection of a minimum of about 10 photons of energy in the microwave region.

With this order of amplifier sensitivity, the onus for noise troubles, very much reduced in magnitude, falls on other parts of the microwave system. Many components, whose contribution to noise could previously be ignored, now must be reconsidered and redesigned. The loss in waveguides or conductors must now be reduced, or they must

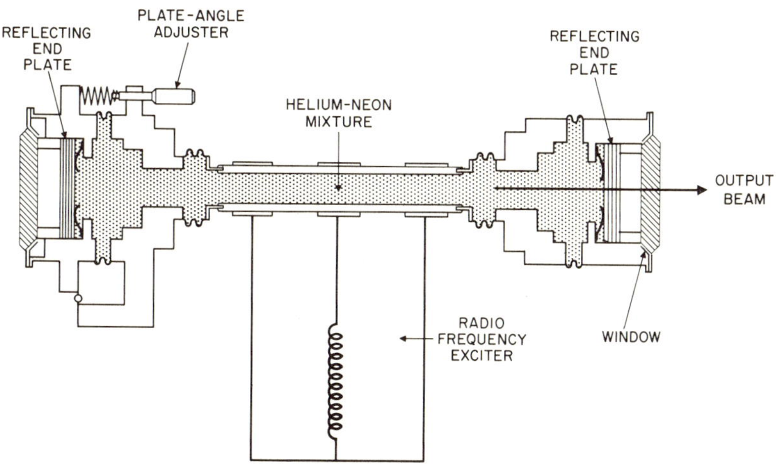

Figure 9. Optical maser as achieved by Javan et al.

Figure 10. Bell Telephone Laboratories low-noise antenna-maser system. (Courtesy of Bell Telephone Laboratories.)

be cooled to very low temperatures. Antennas must discriminate against heat radiation from the ground. The best overall system performance for an entire microwave system so far reported is a noise temperature of 20°K, demonstrated at the Bell Telephone Laboratories with a special antenna (Figure 10). This is an improvement by a factor of about 100 over what could be done before the advent of masers and is possibly close to the practical limit set by the leakage of very small amounts of heat radiation into the system. Yet its performance is still two orders of magnitude poorer than that which might, in principle, be achieved with a maser amplifier, and one can hope for still further improvement.

It is also interesting to note the stimulating effect that maser amplifiers have had on the development of low-noise amplifiers of other types. Since the appearance of masers, the parametric amplifier has received increased attention and its noise temperature has been lowered to about 50°K, a ten- to twentyfold reduction. Other amplifiers have also been appreciably improved. It is likely that an important factor in bringing about these improvements was the awareness that we need no longer accept levels of noise which previously seemed irreducible.

It is partly because masers are so free of noise that maser oscillators (which are simply amplifiers with sufficient positive feedback to make them oscillate) are of such high quality and make good atomic clocks. Another reason is the fact that they oscillate only within the response width of the particular atomic or molecular resonance chosen. Some of these resonances are exceedingly sharp and stable in frequency.

There are other types of very good atomic clocks; in fact, a system using a cesium beam with no maser action at all, developed under the direction of Professor Zacharias of M.I.T., has been remarkably successful and is widely used at present. However, the clock which seems to promise the highest accuracy is a direct descendant of the original ammonia maser. It is a maser oscillator using a beam of hydrogen atoms, now under development in Professor Ramsey's laboratory at Harvard.[17] An early design is shown in Figure 11. This device has not yet been well tested, but it is hoped that constancy over a long period of time of possibly 1 part in 10^{13} can eventually be obtained.

Optical and infrared masers even seem likely to have considerably more importance than other types. Optical masers are probably not generally useful as detectors of light waves, because we already have detectors which are sensitive enough to respond to only a few quanta of light. These are photocells or, at still shorter wavelengths, X-ray and gamma-ray counters. But as oscillators operating in the optical region, masers give us potentialities which were hardly dreamed of before their advent.

Until very recently, all sources of light have been essentially like hot,

Figure 11. Hydrogen maser clock at Harvard. (Courtesy of Harvard University.)

glowing bodies. The intensity of light emitted by a hot source, such as an incandescent lamp, is strictly limited by the laws of black-body radiation; that is, at a certain temperature, no source can emit more than a certain quantity of energy per square centimeter, within a certain band of wavelengths or frequencies. It is true that this energy increases as the temperature of the body increases, but the intensities which we have been able to achieve are limited by temperatures of a few tens of thousands of degrees attainable in ordinary materials. This has seemed a natural limit for so long a time that we have grown to accept it. But imagine where radio technology would have been if we were limited to the radio waves produced by heat radiation. The uses of radio waves would have been sadly restricted. Fortunately, radio engineers have for a long time been blessed with coherent oscillators which could produce intensities of radiation enormously larger than those from thermal sources.

Optical and infrared masers have provided coherent oscillators in this new frequency range which have suddenly increased enormously the amount of energy achievable per unit bandwidth, thus allowing a technology rather more like that of electronics than that of classical optics. Intensities of these oscillators are so high that temperatures of the order of 10^{20} degrees would be required in order to reproduce them by heat radiation. This increase in energy per unit bandwidth per unit area is fundamental to the new freedom yielded by masers operating in the optical region. Concomitant with it are many interesting and useful properties.

First, consider monochromaticity, or frequency purity. The frequency of an optical maser is determined in part by the separation between the two mirrors. If this separation remains constant, the only remaining effect which can disturb the frequency of oscillation is an occasional quantum of energy spontaneously emitted by some atom in precisely the same direction as the beam. In typical masers, one can show that this will produce random fluctuations of the oscillation frequency of the order of 1/1,000, or a cycle per second. Since the light is oscillating at the rate of about 10^{15} cps, this corresponds to fractional frequency changes of 1 part in 10^{18}.

In a maser oscillator of the type invented by Javan, the essential parts consist of a gaseous discharge of helium and neon and plane parallel plates at either end of the discharge. Between these plates, a standing wave oscillates stably at a frequency near 3×10^{14} cps. If one of the mirrors is only partially silvered, some infrared light leaks through in the form of a directional beam. Experiments show that over short periods of time the oscillations represent a pure frequency to at least 1 cps. Over longer periods of time, acoustic vibrations and temperature effects move one mirror with respect to the other and produce larger slow variations of frequency. For example, if one mirror moves by 1 angstrom unit (10^{-8} cm) in distance, the frequency changes by a few tens of kilocycles, or about 1 part in 10^{10}. This provides an exceedingly sensitive detection of small motions.

The radiation emitted by such a maser is coherent not only in time, but also in space; that is, over the surface of the mirror, the light wave is oscillating almost entirely in phase. As a result, the light leaking through the mirror is highly directive and represents a much closer approximation to a plane wave than has previously been achieved. The directivity of the beam is limited only by the fundamental laws of diffraction; that is, the angular spread of the beam is given approximately by λ/D, where D is the diameter of the mirror aperture and λ is the wavelength of the radiation. For our particular case, this angular divergence of the beam is near 10^{-4} radian, about one-hundredth the divergence of sunlight or of a very good searchlight.

Moreover, still greater directivity can be achieved by using a lens to converge the beam. It will focus such a beam to a point whose size is limited only by diffraction, and hence to a diameter of about one wavelength of light; that is, the radiation from the maser can be focused so that the entire beam goes through a cross section with a diameter of about one wavelength. Such focusing cannot be achieved with incoherent light, the type we have known before. After the beam diverges again, it can be made parallel by a second lens now very much larger than the first. Its angular spread is still given by λ/D, but now D is the diameter of the larger lens. A perfect 200-in. telescope used as a lens would thus give a directivity to the entire beam which would correspond to an angular spread of about one ten-millionth of a radian. Such a beam in the atmosphere would be somewhat distorted, however, because of variations in the optical properties of the atmosphere, the same variations which produce twinkling and slight apparent motion of the stars.

Light from a good maser can be directed through an optical instrument in such a way that it will be focused on the smallest region which the optical instrument is capable of resolving. Thus, when sent backward through a microscope, the entire energy can be focused on the smallest portion of a biological cell which the microscope is capable of resolving. When focused through a telescope, the beam can be directed to a spot on the moon, for example, which is as small as the smallest region the telescope can resolve when used for normal observation.

Whereas the present gas maser produces a few milliwatts of infrared energy, the ruby optical maser produces many kilowatts of red light in a short pulse. But its beam, though highly directional under normal circumstances, falls short of the theoretical directivity by about a factor of 10, while the gaseous maser approaches very closely to the theoretical limit set by diffraction. On the other hand, it has been shown that light from the ruby maser is almost entirely coherent; hence a suitable optical system can, in principle, focus it to as small a point or as narrow a beam as in the case of the gas maser.

The field strengths produced at the focal point of a beam from an optical maser have not yet been measured, but some calculated numbers are useful as illustrations. One watt of power concentrated at the focal point would have a power density there of about 100 million watts per square centimeter, enormously higher than any other known radiant intensity. Field strengths in the electric wave at the focal point would reach about 1 million volts per centimeter for 1 watt of power, or 100 million volts per centimeter for the 10 kw which a ruby maser normally produces. It must be emphasized that this figure pertains to the field strength due to *light* waves oscillating at a frequency of about 10^{15} cps. The pressure of light at this hypothetical tiny focal point would be near 1,000 atm. Such intensities can easily tear apart atoms and molecules; maser designers have been comparing their achievements by stating how many razor blades one millisecond flash of the light the masers produce will penetrate. A hole blasted through one package of 12 razor blades, including its containing carton, is about par. When focused, the beam is intense enough to evaporate the most refractory materials, as well as razor-blade steel.

Many kinds of optical and infrared masers are now available (Figure 12). New types appear at the rate of about one per month, with an increasing variety of wavelengths and characteristics. However, there are more in the near infrared than in the optical region; masers oscillating at wavelengths shorter than the red are still very rare. The ruby maser, the He-Ne system, as well as some other types, have now been made to oscillate continuously at reduced power. In pulses, masers have produced a power of about 10 megawatts for a small fraction of a microsecond. Frequencies in a given maser have been tuned over a range of about 20 wave numbers, or 600,000 Mcps.

Intensity of light from optical masers is strong enough to produce nonlinear effects in normal optical media and hence to allow the production of harmonics of optical radiation, as demonstrated by Franken, Weinreich, and others at the University of Michigan,[18] who focused a beam of maser light on a piece of quartz, producing the second har-

Material	Wavelength in Angstroms	Operation
Cr^{+++} in Al_2O_3 (ruby)	6,943 7,009 7,041	pulsed and continuous
Sm^{++} in CaF_2 (crystal)	7,082	pulsed
U^{+++} in BaF_2 (crystal)	24,000 27,000	pulsed
Nd^{+++} in $CaWO_4$ (crystal)	10,600	pulsed and continuous
Nd^{+++} in glass	10,600	pulsed
Benzophenone and naphthalene in organic glass	4,700	pulsed
Ne in He (gas discharge)	11,180 11,530 11,600 11,990 12,070	continuous

Figure 12. Characteristics of some maser systems.

monic of the original light at sufficient intensity to be easily detectable. These nonlinearities also allow two frequencies to beat together, so that the sum of the two frequencies is produced.

An older and more familiar device, the photocell, also responds to light in a nonlinear fashion to give the equivalent of rectification. Since the photocell responds to the intensity of light or to the square of the field strength in the light, beams of light from two different optical masers can be sent into the same photocell, where they are mixed; the difference frequency is produced in the electron current of the photocell. This is heterodyne detection at optical frequencies.

We see from the above discussion that there are now available in the optical region coherent amplifiers and oscillators, antennas and methods of transmission, frequency multiplication and beating, rectification and heterodyne detection systems. One may also add to this list filters of various types and designs afforded by sharp optical resonances, as well as attenuators of absorbing material. Thus we now have in optical technology all the normal complement of tools and devices used by the electronics expert.

Although our attention here is primarily on electronic and applied uses of masers, their application to fundamental scientific experiments seems also very promising. Since optical masers afford new orders of precision in measurement of the two fundamental quantities, length and time, we may expect them to allow in due course reexamination of many facets of the physical world with great delicacy and precision. For example, we can now visualize the tools necessary to multiply and divide frequencies in the optical and infrared region, and to amplify the resulting new frequencies so that they can be further multiplied or divided. This encourages one to look forward to the day when radio frequencies, on which our standard of time is based, are multiplied by harmonic generation to frequencies through the infrared and into the optical region, so that we can determine on the basis of our standard of time the frequency of oscillation of optical wavelengths. Since these optical waves give us our standard of length, the product of their length and frequency immediately affords a determination of the velocity of light to a precision as great as that to which the standard of length can be defined. This procedure, in fact, obviates the necessity for separate definitions of standards of length and of time, since the velocity of light can then itself be used as an absolute standard of length.

4

Very preliminary engineering applications of optical and infrared masers have already been made. But it is probably more useful at this stage to examine some of the theoretical possibilities and dream about potential applications than to discuss specific and rudimentary initial work. Let us reconsider some of the classic functions of electromagnetic waves and electronics.

What about communications? Optical frequencies can completely dispel bandwidth problems. One-tenth of one per cent of the optical frequency range corresponds to the entire bandwidth so far used throughout the radio and microwave frequency range. In the optical range there is sufficient bandwidth so that all inhabitants of the earth can talk simultaneously on their own private frequency bands. But in addition, the directivity of optical beams is such that wireless communication can be maintained essentially in complete privacy and without interference from other channels of communication or noise. No one beam gets in the way of another. It is, of course, true and important that optical and infrared beams will not penetrate very far through clouds or fog. This means that for long-distance communication by electromagnetic waves of this frequency one must either use pipes at the earth's surface or go above the clouds. For communications at high altitude and in space, optical and infrared beams may be particularly advantageous since there is no cloud trouble and since the broadcasting or receiving antennas can be only a few inches in diameter and still enormously directive. Nor would the power requirements of such optical masers necessarily be very large. Work is now under way to power an optical-maser oscillator from solar radiation alone.

The distance over which these optical beams can in principle be detected is impressive. This is due in part to their directivity. For illustration, let us look into the future and imagine an optical maser which can deliver 10 kw of power continuously rather than simply in pulses as at present. Let us also assume that this light is directed by a good 200-in. telescope located above the atmosphere so that atmospheric problems produce no unwanted fluctuations in the direction of the beam. At a large distance from the solar system, its beam can look brighter than the sun. This may seem surprising, but remember that its 10 kw of power is highly directed whereas the sun's energy is radiated in all directions. Remember also that the energy in this beam is confined to a very narrow frequency range whereas the sun radiates in a broad frequency range. Relatively simple filters which allow the passage of a narrow frequency range corresponding to that produced by the maser will hence make its beam seem brighter than the sun. The maser light beam could be detected by the naked eye to distances of about $\frac{1}{10}$ light year. Obviously, then, signaling between planets in our own solar system would be easy even by use of the handiest of receivers, the human eye.

With another good telescope as a receiver, the beam could be detected up to distances of about 100 light years. This allows the possibility of interstellar signaling in the optical region. Project OSMA involves an effort to listen for radio waves which may perhaps be directed toward our earth by intelligent beings associated with other stars. Our knowledge of the technology developed by some other form of intelligent life is not sufficiently accurate to ensure that attempts to communicate with us may not be by means of optical masers rather than radio broadcasts. Perhaps we should be looking at neighboring stars with a good telescope and a high-resolution spectrometer for a spectral line much narrower than those emitted by normal atoms, that is, for light which could be produced only by something like a maser.

Although interstellar communication may be a long way off, local communication by optical beams seems immediately practical. In this area, part of the engineering problem is to learn how to modulate and demodulate sufficiently well so that the enormous bandwidth afforded can be properly utilized. Even with much wastage, the additional load of communications which optical waves can carry is impressive.

What about radar? Already, light has been produced at a power level of about 10 megawatts, with a pulse length somewhat shorter than 10^{-7} sec. I see no reason why the pulse length cannot be further shortened to about 10^{-10} sec, with some corresponding increase in the power level. The range of radars using optical or infrared beams could be quite large, and the short pulses would allow a very accurate determination of distances. The directivity of such a radar beam would be several orders of magnitude better than those in current use; hence angular directions could be determined with greatly increased precision. But we are faced again with learning how to utilize this new potential. In terms of the necessary engineering, one important question is whether we can properly take advantage of the extreme directivity of maser beams. Can other system components be made to match their performance, and can we achieve the necessary precise control of optical systems under high power and heating conditions?

Interference patterns produced by optical beams can be used to measure large distances with still greater precision than by light radar pulses. These optical beams are sufficiently monochromatic to allow the formation of interference patterns across distances of many thousands of miles. Hence, for a distant reflector, motion as small as one wavelength of light could be detected easily, assuming that the optical path to it is sufficiently stable, as it would be in space. Do we want measurements of distances to this accuracy, and what use can we make of them? Would we want to use Doppler effects at optical wavelengths to measure the velocity of objects hundreds of miles away with an accuracy of a small fraction of a centimeter per second?

It is clear that there are many practical uses for accurate determination of distances closer at hand. One example is measuring the accuracy of the surface of a large radio telescope dish. A maser light beam reflected from the surface of the dish would produce interference patterns by means of which its shape could be determined to within any precision desired down to about one wavelength of light.

What about power transmission? Transmission of sizable amounts of power over long distances by light beams does not seem too much to hope for. The telescope postulated above to be in orbit above the atmosphere could concentrate the 10 kw of power of a maser beam into a spot about 6 in. in diameter 1,000 miles away. A flippant thought is that the spot is about the right size and has plenty of power to heat some spaceman's morning coffee, assuming he knows just where to put the coffee pot. Perhaps in the future satellites will be supplied with power by means of optical beams—perhaps even pushed into position by light pressure.

A good telescope placed on the moon could direct a maser beam to a particular spot on earth about the size of a football field. Is it only fantasy to ask whether this will be a future form of night illumination which can be beamed to various spots in accordance with local needs? And shall we have to face commercialization in the form of bright-light advertisements on the moon?

But back on earth there are mundane applications of intense or of monochromatic and coherent beams of light which are more immediate, and probably more important. These include etching and welding, which may be carried out with great precision and under chemically pure conditions, with surfaces touched only by light; surgery or microsurgery performed with great speed and exactness on any visible surface; increased intensity of illumination for photomicroscopy under high magnification; extremely precise measurements of distances and angles in surveying, in determination of small distortions, and in very precise control of machines.

There is little doubt that we shall see many of these applications. But precisely which ones will be most successful and most exciting, and what additional new concepts and techniques will emerge, cannot be so easily foreseen. The full nature, or even all the broad fields of application, of masers and other quantum electronics devices are surely not yet evident. In the region between microwave and infrared frequencies, we still do not have sources of electromagnetic waves of sufficient intensity for proper scientific or technical work. But it is quite easy, now, to be convinced that masers will yield such sources. Very high frequency acoustic oscillators, operating on maserlike mechanisms, can also be expected. Masers should operate in the ultraviolet region and at still shorter wavelengths, perhaps approaching the soft X-ray region, where one can show that present ideas for masers clearly will no longer work. Coherent oscillators at wavelengths as short as those of X rays will require some new discovery or invention.

Certain aspects of the future development of masers and quantum electronics can be envisaged, including a number of interesting ones which space does not allow to be discussed here. But as a humbling further reminder of the unpredictable and frequently happy surprises of research, we have only to ask what research director would have been wild enough, a decade ago, to set about the development of a very low noise amplifier, or a more accurate clock, or a very intense light source by asking his research staff to study the absorption of microwaves in gases or in paramagnetic materials at low temperatures.

For those interested in electronics, optical and infrared masers seem likely to have a particularly powerful effect in the future. They will take electrical engineers deeper into spectroscopy, very high precision optics, and the strange world of quantum mechanics. This touches on what is perhaps one of the most important aspects of the maser, recently described with some humor by an editorial writer.[19] "We propose a new name . . . to include all those new concepts and devices that allow, and even force, the scientist-engineer to break into a new field of science and technology and of thought. We propose the name gaser, coined by the well-established method from growth *a*cceleration by the *s*timulated *e*xcitation of *r*esearchers. We suggest that the maser is a gaser."

It is a pleasure to acknowledge the good counsel and considerable help given me in preparation of all phases of this article by Dr. James W. Meyer of the Lincoln Laboratory.

REFERENCES

1. "Peacetime Uses of Outer Space," S. Ramo (ed.), p. 46, McGraw-Hill, New York, 1961.
2. J. A. Giordmaine, L. E. Alsop, C. H. Mayer, and C. H. Townes, *Proc. IRE*, vol. 47, p. 1062, 1959.
3. C. E. Cleeton and N. H. Williams, *Phys. Rev.* vol. 45, p. 234, 1934.
4. C. H. Townes, *Bell Telephone Lab.* Memo., 1945.
5. E. M. Purcell and R. V. Pound, *Phys. Rev.*, vol. 81, p. 279, 1951.
6. It is probably useful here to provide a short summary of nascent ideas and early events connected with the maser. In a recent paper, F. G. Houtermans (*Helv. Phys. Acta*, vol. 33, no. 8, pp. 933–940, 1960) recalls having realized in the mid-thirties the possibility of exponential build-up of photons by stimulated emission, or a "photon avalanche," and describes the molecular system in which he was interested at the time. I understand that V. I. Fabricant in the U.S.S.R. included a few lines on "negative absorption" in a thesis in 1940 (not so far available in the United States) and applied for a patent in this field in 1951, which was turned down at the time but was issued in 1959 (presumably in modified form). W. E. Lamb and R. C. Retherford discussed conditions for "negative absorption" in a gas clearly, though briefly, in 1950 (*Phys. Rev.*, vol. 79, p. 549). There may be still other references of this type in the three decades of spectroscopic work after 1920 which have been overlooked so far. In 1951 Purcell and Pound demonstrated inversion of population and "negative absorption" of nuclear resonances, without net gain (*Phys. Rev.*, vol. 81, p. 279). Inversion of population had already been demonstrated in 1946 by F. Bloch, W. W. Hansen, and M. Packard with adiabatic fast-passage techniques, although the experiment actually measured reversed dispersion rather than "negative absorption" (*Phys. Rev.*, vol. 70, p. 474). A molecular oscillator (the beam-type maser) for obtaining net gain and oscillation seems to have been first discussed by A. H. Nethercot on behalf of the author at a Symposium on Submillimeter Waves at the University of Illinois in May 1951.

Design of this device was outlined in the Columbia Radiation Laboratory Quarterly Progress Report of December 1951, and experimental work reported in subsequent issues. Successful oscillation of the maser was reported by J. P. Gordon, H. J. Zeiger, and C. H. Townes in 1954 (*Phys. Rev.*, vol. 95, p. 282). N. G. Basov and A. M. Prokhorov, of the Lebedev Institute in Moscow, mention (*Doklady Akad. Nauk S.S.S.R.*, vol. 101, p. 47, 1955) having pointed out "the theoretical possibility of constructing a molecular generator" at an All-Union Conference on Radio-Spectroscopy in May 1952. In 1954, they discussed (*J. Exptl. Theoret. Phys. U.S.S.R.*, vol. 27, p. 431) in some detail the theoretical design of a beam-type device rather similar to that separately developed at Columbia University. J. Weber, of the University of Maryland, considered the general possibility of utilizing stimulated emission for amplification, and a particular system using ammonia gas (without feedback), which he discussed at the Institute of Radio Engineers Electron Tube Conference in Ottawa, June 1952. His paper in 1953 (*IRE Trans. on Electron Devices,* vol. 3, p. 1) considered in some detail a number of the principles involved.

7. C. H. Townes, Sigma Xi Lecture, 1951, *Am. Scientist,* vol. 40, p. 270, 1952.
8. J. Combrisson, A. Honig, and C. H. Townes, *Compt. rend.*, vol. 242, p. 2451, 1956.
9. G. Feher, J. P. Gordon, E. Buehler, E. A. Gere, and C. D. Thurmond, *Phys. Rev.*, vol. 109, p. 221, 1958.
10. N. Bloembergen, *Phys. Rev.*, vol. 104, p. 324, 1956.
11. H. E. D. Scovil, G. Feher, and H. Seidel, *Phys. Rev.*, vol. 105, p. 762, 1957.
12. A. L. Schawlow and C. H. Townes, *Phys. Rev.*, vol. 112, p. 1940, 1958.
13. T. H. Maiman, *Nature,* vol. 187, p. 493, 1960.
14. R. J. Collins, D. F. Nelson, A. L. Schawlow, W. Bond, C. G. B. Garrett, and W. Kaiser, *Phys. Rev. Letters,* vol. 5, p. 303 (1960).
15. A. Javan, *Phys. Rev. Letters,* vol. 3, p. 518, 1959.
16. A. Javan, W. R. Bennett, and D. R. Herriott, *Phys. Rev. Letters,* vol. 3, p. 106, 1961.
17. H. M. Goldenberg, D. Kleppner, and N. F. Ramsey, *Phys. Rev. Letters,* vol. 5, p. 361, 1960.
18. P. A. Franken, A. E. Hill, C. W. Peters, and G. Weinreich, *Phys. Rev. Letters,* vol. 7, p. 118, 1961.
19. M. C. Peace, *Microwave J.*, January, 1962. We have taken the liberty of changing, in his text, the word "laser" to "maser."

John R. Pierce, Executive Director, Research, Communications Principles Division, Bell Telephone Laboratories, Murray Hill, New Jersey, has made fundamental contributions to the theory and design of electronic devices. The work of his division includes research on passive and active satellites for transoceanic communications.

CHAPTER 9

SATELLITE RELAYS

So much work is now being done on satellite communication that it has become very difficult to talk briefly and intelligibly about technical details. One has the choice either of discussing some particular point at length or of being very general. Part of this discussion will be a general consideration of the two technical aspects of satellite relays: the art of space and the art of electronics and communication. But generalities are apt to be meaningless unless they are referred to specific things. For this reason, broad principles will be considered in relation to the experiments with Echo and the Telstar active satellite, the two Bell Laboratories ventures into the field of satellite communication.

1

The art of electrical communication has a long history of accomplishment, starting with the invention of the telegraph in 1832. Yet those concerned with electrical communication see no end to fresh, new devices, techniques, and systems. Certainly, advances in electrical communication will keep step with advances in even as rapidly changing a field as space.

By contrast, the art of space seems entirely new to the casual observer. Certainly, it still has the fallibility of infancy, and it is without great practical application save in favoring offense over defense in nuclear war. Yet, new as it seems, it would be a mistake to think of the art of space as entirely, or even largely, new.

One of the most vital aspects of the art of space is our understanding of, and ability to compute, the trajectories of rockets and the orbits of satellites. Yet Isaac Newton's enunciation of the laws of motion, his discovery of the law of gravitation, and his invention of the method of fluxions, which we now call the calculus, came in the seventeenth century, not the twentieth. Work under space contracts has no doubt refined mathematical methods, but we owe the greater part of our ability to compute trajectories and orbits to the venerable science of celestial mechanics.

Celestial mechanics has benefited in recent years from a powerful new device, the electronic digital computer. Space science would be almost hopeless without digital computers; yet new as digital computers are, their origin lies outside of the field of space. It goes back to the work on relay computers by Aiken and his associates at Harvard; to Stibitz, Andrews, Williams, and others at the Bell Laboratories in the forties; to the storage of programs in computer memory by Von Neumann. The digital computer was first made into an electronic device, the Eniac, by Eckert and Mauchly in 1945. The fact that we have powerful electronic computers available for space uses is due to the diligence and enterprise of several companies, whose chief interests are other than space.

Computers, calculus, and the laws of mechanics and gravitation are necessary for guidance, but they are not sufficient. One other element is inertial guidance, which we owe to years of patient endeavor by men, including C. Stark Draper, who were motivated, in the production of gyroscopes of exquisite and almost unbelievable precision, by problems of aerial navigation rather than space flight.

Besides inertial navigation, space flight relies on radio tracking and control. Here it has drawn on a tremendous amount of microwave and mathematical art, some preceding World War II, and more an outgrowth of wartime and postwar radar and microwave communication. This is an art complex beyond any brief description—and there it was, waiting to be applied in the field of space. To these arts of microwave radar and communication, we should add other electronic arts, including the transistor, to amplify and control with high reliability and little expenditure of power, and the solar cell and the nickel-cadmium storage cell, to supply those low levels of power in satellites. We should also add the servomechanism art, which was brought to the perfection required in space by both prewar and postwar work in other areas of application.

Just as the art of space has drawn its knowledge and practice of guidance and control from external sources, so too, a satisfactory structure for space vehicles was possible only through art developed in connection with airplanes. Partly, this provided special alloys, such as duralumin, which was invented prior to World War I. Partly, it provided means of analysis and design and means of assembly by welding and riveting. It is no mere coincidence that most rocket vehicles are designed and built by airframe manufacturers.

The rocket motor itself, first put into practical operation in the German V-2, has a novelty which some other aspects of space art lack. Yet, in understanding the conversion of chemical energy into propulsive force, we must rely on thermodynamics, a nineteenth-century science which came into being as a result of the invention of the steam engine early in the eighteenth century and of the knowledge of the kinetic theory of gases, which we owe to Maxwell and Boltzmann. Furthermore, the production and utilization of fuels draw on both the petroleum industry and on the industrial liquefaction of gases.

Some aspects of the space art, such as ablation cooling of nose cones, seem truly new. But when we consider most major aspects, such as trajectories and orbits, guidance and control, the structure and fabrication of space vehicles, and rocket motors, we are forced to conclude that space art is an adaptation and an exploitation of the world's science and technology rather than a truly independent science and art. Despite the huge sums and great effort that have been spent on space, huger sums and greater effort have been spent over many, many years in creating the science and technology which have made rockets and satellites possible. It would be perfectly feasible to control and restrict the exploitation of space art in other fields of technology, but any limitation on the use, in space, of our general pool of science and technology would render that new but dependent art not difficult, but impossible.

Truly, it is wrong and unfruitful to compartmentalize science and technology. Rapid progress is possible only if various aspects of science and technology are allowed to interact freely, so that old problems can be solved and new potentialities exploited. Indeed, aspects of science and technology seem most set apart when they are newest and feeblest, as were the automobile and the airplane when they were sportsmen's toys. Today, cars and planes have insinuated themselves into all sorts of technological niches. As they grew stronger, they became parts of our lives, and they have more than made up in usefulness what they have lost in glamor.

2

We may ask how an art, whose objectives seem as strange and remote as space, can outgrow its childhood and repay its debt to the rest of technology. Space travel, except for nuclear weapons, seems remote from practical aims. Weather forecasting offers a possibility of useful application. But it is chiefly in the field of electrical communication that we can look for an early, effective application of the art of space.

This is so for two reasons. One is that the users of the world's 142 million telephones can exploit and support electrical communications on a scale commensurate with the large capabilities and cost of a satellite communication system. The other is that a satellite communication system promises to provide service at a lower cost than other alternatives of comparable quality.

The fact that in many nations there are many telephone users who are potential users of satellite communication is of great importance. This is of course not so of all countries. It will take years to create such a situation in many new or backward nations, where adequate internal electrical communication must first be provided, and society must so reshape its customs as to need and to rely on electrical communication.

Even in countries with the largest number of telephones, widespread application was slow in developing. As in the case of the space art, telephony became possible only through a long accumulation of knowledge and understanding of electrical phenomena, and its exploitation and improvement have been dependent on the same mathematics which has been essential in space. When the telephone was invented by Alexander Graham Bell in 1875, it was crude and weak and enabled people to talk over short distances only. The invention and development of the carbon microphone and the induction coil in 1877–1890 provided amplification which greatly increased the quality and range of the telephone. By 1913, loading had made telephony possible over distances as great as 2,000 miles. The use of triode vacuum tubes brought transcontinental telephony into being in 1915. In 1927, overseas telephone service was first provided by means of long-wave radio; in 1929, short-wave radio was put into use, and telephony became an international means of communication. In 1956, the first transoceanic telephone cable, which provides higher quality and greater reliability than short-wave radio, was placed in service.

More than advances in capability were needed, however. In 1920, the charge for a 3-minute telephone call from New York to San Francisco was $16.50; today the charge is $2.25. The reduction of long-distance rates, in the face of a steady rise in prices, has been essential to the widespread telephone service we now have. This reduction has been brought about through the application of new principles and new devices in the transmission of speech.

The transmission of many messages over one pair of wires—4 over an open-wire circuit in 1918 and 1,800 over a coaxial cable today—has been essential in this cost reduction. So has the use of microwave radio, which was put into transcontinental service in 1951. These advances required the use of new vacuum tubes, both conventional and microwave, quartz-crystal stabilized oscillators, crystal filters, and other new devices, and of new principles such as negative feedback and frequency modulation. Today the transistor and pulse code modulation promise a reduction in the cost of short circuits, and millimeter waves and waveguide, a reduction in the cost of long circuits.

Thus telephony has attained the place it has in our lives partly because it enables us to reach almost any point in the country and many places in the world, but partly also because advances in the electronic art have reduced the cost of telephone transmission.

Many of these advances, such as the transistor and negative feedback, have come from research carried out in the telephone industry, but others, such as the triode, which ushered in the age of electronics, the klystron, and the traveling-wave tube, were invented outside of the telephone industry, to be applied therein.

While the art and science of telephonic communication are certainly closer to self-sufficiency than those of space, telephony too is merely one aspect of our technology and science. It is not self-sufficient, and it would be hard put to survive if denied the use of inventions and discoveries, that is, of technology and science in general.

For telephony has its unsolved problems. One is that of providing circuits with bandwidth large enough for television transmission, at a cost low enough to permit extensive use of television for other than broadcast purposes. The telephone companies of various countries of the world have admirable facilities for the transmission of television programs, but the costs are such as to limit severely other uses, such as person-to-person or conference television.

Another problem of telephony is that of providing adequate electrical communication across oceans. Short-wave radio is inadequate in both the quantity and the quality of communication it provides. The cost of a submarine cable is comparable to that of a transcontinental microwave radio relay system with more than ten times the cable capacity. And the submarine cables of the present and near future are too narrow-band to transmit real-time television at all.

3

It is this problem of transoceanic communication that can bring together two aspects of science and technology, space and electrical communication, to the benefit of both. Space art can provide, by means of a satellite, a relay point visible from both sides of an ocean, and communication art can provide equipment to beam a microwave signal to the satellite from the ground, to receive, amplify, and retransmit it at the satellite, and to receive the signal from the satellite at the far side of the ocean.

What is necessary if this is to be accomplished?

One essential is a great deal of work, comprising both research and development. Laudable as past accomplishments have been, they have been no more than first steps toward a practical operating satellite communication system. There is reason to be pleased that, after low-altitude inflation tests, Echo I was successfully launched by NASA on the second try, on August 13, 1960. The tracking, transmitting, and receiving equipment at the Jet Propulsion Laboratory and Holmdel worked so satisfactorily that transcontinental transmission of good quality was achieved on the first pass. It should be realized, however, how uncertain the art of space is, and how even the best-planned experiments, such as the first West Ford launching, can go awry.

Yet reliability is of the utmost importance in any practical space communication system. People are used to reliable communication; they will not, nor should they, tolerate a service with appreciable failures. This makes it very important to consider how, in the face of the uncertainties of the space art, we can advance most quickly toward a practical satellite communication system.

It seems to me that, in order to establish quickly the required new standard of reliability and life which is essential to satellite communication, two kinds of activities are necessary:

1. Relatively simple projects, which can be successful and useful in the near future, must be carried out.
2. Research and development necessary to the successful undertaking of more ambitious projects in the more distant future must be carried out.

Fortunately, these goals are completely consistent with one another. Both call for aiming at the *early* attainment of *specific goals,* each of which is a step beyond our present achievement or resources. Such steps will in themselves be worthy technical achievements, valuable in the eyes of the world; they can take us most quickly and surely to attainments beyond any that we can now hope to achieve on the basis of our present imperfect art and knowledge.

I believe that we can learn something valuable from past experience with SCORE, Echo, Courier, and West Ford.

SCORE was a very early demonstration that satellite communication is possible. As such, it was worthwhile even though its life was short.

With a great deal of good, hard work, Echo was just within the limits of the art. It demonstrated two-way transcontinental telephone communication. It showed that flimsy, inflatable structures have a surprisingly long life in space. It brought new products of the highly developed communication art, such as low-noise maser receivers, horn-reflector antennas, and FM-with-feedback receivers, into use in the field of space for the first time; all these will be important in satellite communication systems. It demonstrated the feasibility of tracking satellites automatically on the basis of orbital data.

The 18-day life of Courier was a blow to satellite communication. Courier was a very complicated satellite. Either the goal set was too difficult, or the technical work was not well enough done. It seems likely that the quality of electronic devices and circuits used was not equal to the task.

The failure of the first attempted Echo launching on May 13, 1960, and the failure of an attempt to launch an experimental dipole belt in connection with project West Ford on October 21, 1961, illustrate the uncertainties of the space art. Each experiment was well thought out, and each failed because of a failure of space art rather than communication art.

Communication satellite launchings for 1962 include further balloon launchings, by NASA, of 135-foot spheres with some rigidity; placing Echo II in an 800-mile orbit; and, in Project Rebound, placing three spheres in 1,700-mile orbits. The Bell System Telstar satellite, to be launched by NASA at Bell System expense, and NASA's Project Relay satellite, to be built by RCA, are simple, spin-oriented satellites with essentially nondirectional antennas. They will be placed in elliptical orbits with a perigee of a few hundred miles and an apogee of a few thousand miles by Thor Delta vehicles. They will demonstrate transatlantic transmission of telephone and television signals and make radiation measurements as well. A ground station for use with these satellites has been built at Andover, Maine, by the Bell System. The British Post Office is building a ground station in southwest England, and the French Ministry of Telecommunications is building a ground station in Brittany. NASA's Syncom satellite, to be built by Hughes, is a very simple 24-hour satellite of limited bandwidth. It will be spun about an axis normal to the plane of the orbit. The antenna will be directive in a plane through the spin axis, but not in a plane through the equator.

A 1962 launching is also planned as a part of the Army's Advent program. This will be a medium-altitude test launching, in which attitude control, station keeping, and communication equipment will be tried out.

The goal of Advent has been a highly elaborate satellite communication system designed to meet detailed and stringent military requirements. This calls for a very heavy payload, oriented and kept accurately in position in a 22,300-mile high, 24-hour "stationary" orbit. So far, none of the difficult features of Advent have been brought to a realistic test. Instead of having demonstrated a sequence of advances in the art, the program remains in question. If all the features of Advent work well and for several years, a sudden and tremendous step forward will have been taken. If one or more features fail, and especially if they fail unaccountably, little progress will have been demonstrated.

Advent is a bold adventure indeed, but there are in addition several simple, experimental satellite communication projects that are near to the limits of the art, but should not be beyond it. A description of one of these, the Telstar project, will show how the work on Echo laid a groundwork for the Telstar experiment, just as the Telstar experiment can provide a foundation for future satellite communication systems.

4

The Telstar satellite will be launched in an elliptical orbit, inclined 45° with respect to the plane of the equator, with a perigee of about 500 nautical miles and an apogee of about 3,000 nautical miles. The Thor Delta vehicle, which will launch the satellite, cannot place the satellite in a circular orbit high enough to be useful for transatlantic communication experiments. The elliptical orbit also has the advantage that it passes through the Van Allen belt and allows measurement of the radiation.

The apogee at first will occur near the equator on the south-to-north crossing, while the perigee will occur near the equator on the north-to-south crossing. Over a period of months the orbit will precess, so that the apogee will occur over the Northern Hemisphere. The most favorable condition will occur about 8 weeks after launch. At this time, the satellite will be visible at Maine for more than 4 hours a day, on the average, and between Maine and Europe for more than an hour a day. During this period, the satellite will be visible between Maine and Europe for as long as 40 minutes at a time, and on 30 per cent of the orbital passes it will be visible for more than a half hour. When,

20 weeks after launch, the apogee is in the Southern Hemisphere, there will be essentially no mutual visibility between Maine and Europe, but communication will be possible between Maine and the Holmdel ground station first used with Echo I.

The Telstar satellite (Figure 1) is roughly spherical in shape with 72 flat facets. It is $34\frac{1}{2}$ inches in diameter and weighs about 180 pounds. The metal framework is made of magnesium, while the shell is aluminum-coated with magnesium oxide to achieve the desired temperature through control of radiation. Solar cells are mounted on most facets. These, together with nickel-cadmium storage cells, will provide power for continuous operation of command equipment and operation on command of the microwave transmitter. The satellite will spin at 180 rpm.

Figure 1. Telstar satellite. (Courtesy of Bell Telephone Laboratories.)

The apertures around the equator of the satellite serve as circularly polarized microwave receiving and transmitting antennas, almost omnidirectional, but having nulls in a cone about the axis of spin. The erectable antenna serves for command and telemetry.

The microwave transmitter makes use of a traveling-wave tube (Figure 2) based closely on prototypes which have been tested for 4 years without failure. The total radiated power is $2\frac{1}{4}$ watts at a center frequency of 4,170 Mcps. In addition, a CW signal at the local oscillator frequency (4,080 Mcps), of a few tens of milliwatts power, is radiated for tracking. The microwave frequency of transmission to the satellite is 6,390 Mcps. The bandwidth of the satellite repeater is 50 Mcps.

In addition to communication equipment, the satellite provides telem-

Figure 2. Traveling-wave tube for Telstar I. (Courtesy of Bell Telephone Laboratories.)

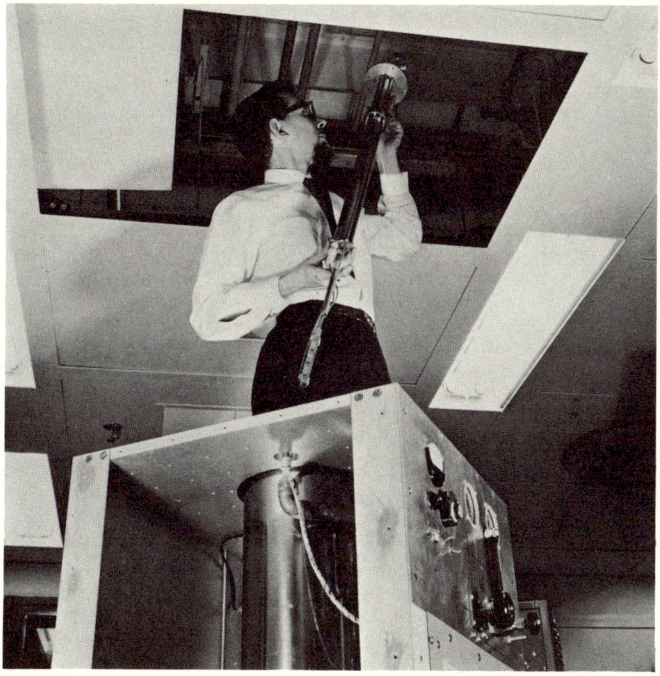

Figure 3. Maser for Telstar reception. (Courtesy of Bell Telephone Laboratories.)

etry channels for reporting 115 items to the ground, at a frequency of 136 Mcps. These items include operating information on communication equipment, radiation measurements, and data on test solar cells and transistors.

The Telstar satellite is capable of relaying a one-way signal of television bandwidth between Andover and Holmdel, the path over which the first tests will be made, or between Andover and Europe. It can also be used to provide a number of two-way voice channels, the single amplifier being used for two directions of transmission simultaneously on different microwave carrier frequencies.

Such communication is possible only through the use of maser receivers and large antennas. Figure 3 shows the maser amplifier used; this has a noise temperature of about 5°K. Because side and back lobes of most antennas receive noise from the ground, the maser will be used with a horn-reflector antenna which has very low side and back lobes. In connection with the 20-foot (400 square feet) Holmdel horn-reflector antenna (originally used with Echo I) (Figure 4), this will result in a

Figure 4. Horn-reflector antenna at Holmdel. (Courtesy of Bell Telephone Laboratories.)

receiver with a noise temperature of about 10°K. Because the 3,600-square-foot Andover antenna (Figure 5) is surrounded by a 210-foot radome to allow operation in high winds, the noise temperature will be perhaps 20° higher than that of the Holmdel antenna.

Telstar will be tracked by means of a technique first employed in connection with Echo I—tracking according to predicted orbital position. The acquisition and tracking equipment is, however, much more elaborate than that used with Echo I. A command tracker, having a 20° beam width, will pick up Telstar's 136-Mcps beacon. The microwave transmitter can then be turned on by a 120-Mcps command signal.

When Telstar has been located and the microwave beacon turned on, an 8-foot precision tracking antenna will obtain very accurate angle information. From this, the satellite orbit will be computed and tapes for pointing the large horn-reflector antenna will be prepared by means of one of two IBM 1620 computers.

From the tapes, pointing information at 4-second intervals will be fed to the drive of the horn-reflector antenna. This will enable the horn reflector to acquire the satellite. Any small errors in pointing can be corrected by self-tracking equipment which makes use of the 4,080-Mcps beacon.

In addition to communication command and tracking equipment, the Andover site (Figure 6) is equipped with a variety of other communication equipment, including tape equipment for recording and originating TV signals. A microwave system will supply two two-way TV channels between Andover and broadband common-carrier circuits at Portland, and wire circuits will supply 46 voice-frequency circuits.

There is more equipment at Andover than would be essential in an operating satellite terminal. Every effort has been made to provide duplication and backup to a degree that will assure success of the Telstar experiment.

5

This, then, is the Telstar part of the 1962 satellite communication experiments. Together with Relay, Syncom, Echo II, Rebound, West Ford, and perhaps Advent, it will lay a groundwork for useful satellite communication systems. What further research and development are needed to make satellite communication feasible, attractive, and economical?

One very important requirement is, of course, a reliable vehicle. NASA's Scout appears to be reliable and economical, but it is too small to launch communication satellites.

Thor is a very reliable vehicle, but in its present form and in present configurations (Thor Delta, for instance) it is too small to launch communication satellites in high enough orbits for practical use.

Figure 5. Horn-reflector antenna at Andover, Maine. (Courtesy of Bell Telephone Laboratories.)

Figure 6. Ground station site, Andover, Maine. (Courtesy of Bell Telephone Laboratories.)

As a part of a multistage vehicle (as Atlas Agena) rather than as a missile, Atlas has not proved comparable in reliability to Thor. There is no reason to believe that results will be better with Atlas Centaur. A vehicle of about the Atlas capability will be necessary for test launchings during the development of communication satellites, and it could be used for launching satellites for an operating system. Either Atlas must be made reliable for such use, or some other vehicle must be found.

Eventually, vehicles larger than Atlas will be needed to launch many satellites at once and to launch very large satellites. Here we face again the need for a tremendous advance in reliability. It would be wrong to assume, however, that huge vehicles will displace smaller ones for communication experiments. For many years, reliable vehicles of about the capability of Atlas will be required for research work in connection with new devices and systems.

The ability to control the attitude of satellites, so as to point directional antennas at the earth, is vital to progress in satellite communication. It will be desirable to point solar-cell arrays at the sun, but only when the weight and reliability of the pointing mechanism are a lesser penalty than the weight of the extra solar cells needed in the absence of sun tracking.

Up to the present, the attitude of Atlas nose cones has been controlled for tens of minutes and the attitude of Discoverer satellites has been controlled for a few days, both by active means (spinning wheels or jets).

At low altitudes, the gradient of the earth's gravitational field and the earth's magnetic field can be used to point a satellite at the earth. We do not know whether or not these means will be practical for communication satellites, which must operate at altitudes of perhaps 7,000 miles. If careful analysis indicates their feasibility, careful experiments should be undertaken.

At the same time, an effort should be made to attain and to demonstrate an assured (not average) life of active attitude-control equipment of several years, preferably of 5 to 10 years. The communication art can provide a life as long or longer. It is only by making the art of attitude control similarly reliable that we can be sure that attitude control will not unduly degrade satellite performance.

The problems of station keeping should be tackled with the same aims as in the case of attitude control.

So far, the only actual demonstration of a satellite electronics life which is commensurate with the requirements of satellite communication is that of Vanguard I (over 3 years' life). Nonetheless, we have every reason to believe that good, well-protected solar cells, which are now available, can supply adequate power for many years. This is true even in the heart of the Van Allen belt, if an efficiency of a few per cent is accepted. Similarly, the best storage batteries, electron tubes and transistors, and other solid-state devices that the communication art affords *can* last for many years. Nonetheless, continued work on long-life components and apparatus of new and improved types is of the greatest importance, so that components can always be chosen on the basis of a long background of testing and experience.

The problems of ground equipment are less critical; this art, at its best, is far more advanced than the space art or even the electronic art aboard satellites. Still, the choice we can exercise among various satellite systems and the ease and quality of satellite communication can depend critically on ground equipment.

A low-noise receiver is always better than a noisy receiver; an antenna is always improved by reducing the side lobes and noise. In large central ground stations (though perhaps not in remote peripheral stations), high standards of ground-station performance should be set and attained. This means making masers and low-noise antennas which are economical, practical, and reliable for routine use. In connection with simple, remote terminals, a low-noise receiver which does not involve the liquefaction of gas is desirable. All this calls chiefly for insisting on the best available materials and designs.

It need be neither unduly costly nor difficult to find and track satellites with high accuracy. This is necessary in experimental work and need not be a serious obstacle in the development of a satellite communication system.

At present, the best of the communication art can provide long-life satellite electronic equipment of moderate power. When considerably larger boosters become available, it may be desirable to use higher powers for some communication purposes. Any sort of broadcasting to ordinary ground receivers would require very high powers.

It would be of interest to build and test more powerful power sources and transmitters than available boosters can orbit. It is unreasonable even to consider launching high-power equipment until an assured life of a year, and perhaps more, can be demonstrated on the ground.

It is evident that bringing practical, useful satellite communication into being will be a tremendous task, however one looks at it. It will be possible only because of the resources of two extensive and well-supported arts: the art of space and the art of electrical communication. In their turn, these broader arts owe their existence to the science and technology of the world. For when we examine either the space art or the communication art bit by bit, it is seen that the devices, materials, and processes we find are present in other electrical and mechanical arts, having been drawn from, or having migrated into, these other fields.

This indivisible overlapping of sciences and technologies is at once a most characteristic and a most powerful feature of our age. Space art and the satellites it provides could be useful in telephone systems. Space art has, however, already benefited substantially from electronics and communication art. There is every reason to believe that the development of a reliable and economical satellite communication system would bring further substantial benefits to space art.

INDEX